Before Einstein

ANTHEM NINETEENTH-CENTURY SERIES

The **Anthem Nineteenth-Century Series** incorporates a broad range of titles within the fields of literature and culture, comprising an excellent collection of interdisciplinary academic texts. The series aims to promote the most challenging and original work being undertaken in the field, and encourages an approach that fosters connections between areas including history, science, religion and literary theory. Our titles have earned an excellent reputation for the originality and rigour of their scholarship, and our commitment to high-quality production.

Series Editor
Robert Douglas-Fairhurst – University of Oxford, UK

Editorial Board
Dinah Birch – University of Liverpool, UK
Kirstie Blair – University of Stirling, UK
Archie Burnett – Boston University, USA
Christopher Decker – University of Nevada, USA
Heather Glen – University of Cambridge, UK
Linda K. Hughes – Texas Christian University, USA
Simon J. James – Durham University, UK
Angela Leighton – University of Cambridge, UK
Jo McDonagh – King's College London, UK
Michael O'Neill – Durham University, UK
Seamus Perry – University of Oxford, UK
Clare Pettitt – King's College London, UK
Adrian Poole – University of Cambridge, UK
Jan-Melissa Schramm – University of Cambridge, UK

Before Einstein

The Fourth Dimension in Fin-de-Siècle Literature and Culture

Elizabeth L. Throesch

ANTHEM PRESS

Anthem Press
An imprint of Wimbledon Publishing Company
www.anthempress.com

This edition first published in UK and USA 2019
by ANTHEM PRESS
75–76 Blackfriars Road, London SE1 8HA, UK
or PO Box 9779, London SW19 7ZG, UK
and
244 Madison Ave #116, New York, NY 10016, USA

First published in the UK and USA by Anthem Press 2017

Copyright © Elizabeth L. Throesch 2019

The author asserts the moral right to be identified as the author of this work.

All rights reserved. Without limiting the rights under copyright reserved above,
no part of this publication may be reproduced, stored or introduced into
a retrieval system, or transmitted, in any form or by any means
(electronic, mechanical, photocopying, recording or otherwise),
without the prior written permission of both the copyright
owner and the above publisher of this book.

British Library Cataloguing-in-Publication Data
A catalogue record for this book is available from the British Library.

ISBN-13: 978-1-78527-178-6 (Pbk)
ISBN-10: 1-78527-178-4 (Pbk)

This title is also available as an e-book.

CONTENTS

Acknowledgements vii

Introduction 1

Part I   **READING THE FOURTH DIMENSION**

Chapter One   Imagining 'Something Perfectly New': Problems of Language, Conception and Perception   19

Chapter Two   Constructing the Fourth Dimension: The First Series of the *Scientific Romances*   45

Chapter Three   The Four-Dimensional Self: Personal, Political and Untimely   75

Part II   **READING *THROUGH* THE FOURTH DIMENSION**

Chapter Four   Four-Dimensional Consciousness: The Correspondence between William James and Charles Howard Hinton   107

Chapter Five   H. G. Wells's Four-Dimensional Literary Aesthetic   133

Chapter Six   Exceeding 'the Trap of the Reflexive': Henry James's Dimensions of Consciousness   167

Afterword   195

Bibliography   199

Index   209

ACKNOWLEDGEMENTS

My work on this book would not have been possible without support from many quarters. I first began researching and writing on the topic of the fourth dimension when I was a doctoral candidate at the University of Leeds, and I am thankful to the Overseas Research Students Awards Scheme and the School of English for their assistance in funding my postgraduate work. The archival research undertaken for the book was partly funded by the Brotherton Library Scholarship Fund. Additional support in the form of teaching relief was provided by York St John University.

While I was at the University of Leeds, I benefited from thoughtful and constructive criticism from a number of scholars. I am thankful for the support of the community of scholars at Leeds, particularly Richard Salmon and Bridget Bennett, as well as then members of the postgraduate community who read and commented on my work at various stages: Catherine Bates, Basil Chiasson, Tara Deshpande, Alberto Fernández Carbajal, Daniel Hannah, Caroline Herbert, Kaley Kramer, Jeffrey Orr, Gillian Roberts, Jennifer Sarha and Abigail Ward. In the early days of my doctoral research, Robin Le Poidevin generously agreed to discuss the philosophy of space and time with me, and my external examiner, Ian F. A. Bell, offered friendly advice and support during the examination process and beyond. While I was on research and conference trips, Linda Dalrymple Henderson met with me to share scholarship on Hinton and discuss the fourth dimension. Daniela Bertol kindly granted permission for her art to appear on the cover of this book. I am immensely grateful to Nasser Hussain and Bonnie Latimer, who heroically read, and provided detailed critiques of, the first full draft of this book. Any errors or misreadings are, of course, entirely my own.

Finally, while this work would not have been possible without the moral support of those mentioned above as well as countless others in Leeds, York, Portsmouth, Arkansas, Austin, Pittsburgh and beyond, I am most grateful to Carrie Kifer, whose humour, love and patience make everything possible.

INTRODUCTION

At the conclusion of Henry James's 1896 novel *The Spoils of Poynton*, the protagonist Fleda Vetch struggles to articulate her sense of the 'vivid presence of the artist's idea' she perceives in the maiden-aunt's house at Ricks: 'It's a kind of fourth dimension. It's a presence, a perfume, a touch. It's a soul, a story, a life. There's ever so much more here than you and I!'[1] Her ability to perceive this presence makes her 'the one who knew the most', the central consciousness of this novel whose understanding most closely approaches James's own.[2] In his preface to the New York edition of *The Spoils of Poynton*, James explained that Fleda's 'ingratiating stroke' for him was that 'she would understand'.[3] Fleda refers to this understanding as 'a kind of fourth dimension', a particular choice of phrase that has not gone unnoticed in literary criticism.[4] This is not Einstein's fourth dimension of space-time; Einstein's special theory of relativity was first published in 1905, and his general theory came six years later. In fact, Einstein's ideas did not begin to reach popular audiences until after their confirmation during the solar eclipse of 1919. To which fourth dimension does Fleda refer then, and how does an understanding of this idea contribute to our understanding of this text and others from the same period?

This book provides an answer to these questions by exploring the discourse of hyperspace philosophy and its position within the network of 'new' ideas at the end of the nineteenth century, before the rise of Einstein's popularity in the 1920s. Hyperspace philosophy grew out of the concept of a fourth spatial dimension, an idea that became increasingly debated amongst mathematicians, physicists and philosophers during the 1870s and 1880s in Britain and on the continent, as well as in the United States. English mathematician and hyperspace philosopher Charles Howard Hinton was the chief popularizer of the fourth dimension in Europe and North America and, from 1880 until his death in 1907, he published a number of literary, philosophical and mathematical texts on the subject. The influence of these texts, many of which were published as a series under the title of *Scientific Romances*, ranged surprisingly wide. The present study offers an extended examination of Hinton's work and – crucially – the influence of his ideas on contemporary writers and thinkers.

Increasingly over the past three decades, critical attention has been given to the relevance of pre-Einsteinian theories of the fourth dimension within the shifting aesthetic and cultural values at the turn of the twentieth century; however, the literary value of

---

1 H. James, *The Spoils*, 196.
2 Ibid., 195.
3 H. James, *The Art of the Novel*, 128.
4 See McGurl, *The Novel Art*.

hyperspace philosophy, and particularly of Hinton's *Scientific Romances*, has been largely overlooked.[5] Mention of Hinton is most frequently made in studies of H. G. Wells; Wells employed four-dimensional theory within his early fiction, calling his own proto-science fiction stories 'scientific romances' as well. Similarly, critics have begun to make the connection between Hinton's work and Edwin A. Abbott's 1884 fantasy, *Flatland: A Romance of Many Dimensions*; Rosemary Jann even used the colour plate from Hinton's 1904 book *The Fourth Dimension* for the cover illustration of the Oxford Classics edition of *Flatland*.[6] Over the past decade, a number of literary scholars have offered glimpses of how a careful and nuanced analysis of hyperspace philosophy can inform a more complex understanding of contemporary writers ranging from Henry James to W. E. B. Du Bois to Ezra Pound.[7] Such discussions – while insightful – are scattered and brief, limited to scholarly journal articles or single book chapters. Until now, the most authoritative and sustained exploration of the aesthetic impact of the fourth dimension has been Linda Dalrymple Henderson's groundbreaking 1983 study, *The Fourth Dimension and Non-Euclidean Geometry in Modern Art*. In this work (which was revised and reissued in 2013), and in other shorter publications, Henderson carefully details the occurrence of the phrase 'the fourth dimension' in the writings of well-known authors such as Oscar Wilde, Joseph Conrad, Marcel Proust and Gertrude Stein. Henderson's work, firmly rooted in art historical practice, offers tantalizing glimpses – but falls short of – the literary perspective I offer here.

Before Einstein addresses, for the first time in a full-length study, the cultural life of the fourth dimension at the turn of the century. I begin by tracing the development of spatial theories of the fourth dimension out of the 'new', non-Euclidean geometries of the mid-nineteenth century and proceed to analyze Hinton's role as four-dimensional theorist and popularizer of hyperspace philosophy. I examine his *Scientific Romances* in detail, not simply as documents of interest for historians of science and ideas, but for their intrinsic literary value as well.

Charles Howard Hinton (1853–1907)

When introducing his translation of three of Hinton's romances as part of his *Biblioteca de Babel* series, Jorge Luis Borges began,

> If I am not mistaken Edith Sitwell is the author of a book entitled *The English Eccentrics*. No one has more right to appear in its hypothetical pages than Charles Howard Hinton. Others seek and achieve notoriety; Hinton has achieved almost total obscurity.[8]

---

5 One exception is Rucker's publication, *Speculations on the Fourth Dimension*.
6 See Abbott, *Flatland*.
7 See, for example, McGurl, *The Novel Art*; Bell and Lland, 'Silence and Solidity in Early Anglo-American Modernism', Parts I and II; and Bentley, 'The Fourth Dimension: Kinlessness and African American Narrative'.
8 Borges, *The Total Library*, 508–09.

Borges is correct: although recently there has been renewed interest in Hinton's work, by the 1940s he was nearly forgotten.[9] His obscurity was partly historical accident – his theory of the fourth dimension was overshadowed by Einstein and Minkowski's work – and partly by design. Personal scandal led to Hinton's disappearance from the British intellectual scene in 1888. However, by the early 1880s, Hinton's career was off to a promising start. The son of fashionable Harley Street aural surgeon and mystic James Hinton (1822–1875), Charles Howard Hinton was educated at Rugby and then Oxford.[10] James Hinton was a founding member of the Metaphysical Society and had his own circle of disciples, including Havelock Ellis. James Hinton's influence – particularly amongst his circle of acquaintances including Ruskin, Tennyson, George Eliot and the family of late mathematician George Boole – was no doubt beneficial to a son who was just beginning to make a name for himself. After graduating from Balliol College, Charles Howard Hinton edited his father's posthumous *Chapters on the Art of Thinking* (1879) and, in 1880, he married Mary Ellen Boole (daughter of the mathematician George Boole). After accepting the position of Science Master at Uppingham School in 1881, Hinton saw some early success in publishing his own work: his early *Scientific Romances* and his 1884 textbook, *Science Note-book*, were reviewed favourably by *Nature*, *Mind* and other periodicals.

However, after his father's death in 1875, rumours of the elder Hinton's sexual improprieties continued to spread; a proponent of 'free-love', James Hinton had died unexpectedly after a period of mental illness that, according to some, looked suspiciously like late-stage syphilis.[11] To make matters worse, in 1883, three years after his marriage to Mary Boole, the younger Hinton bigamously wed his long-standing mistress, Maude Florence, doing so under the pseudonym John Weldon. Maude was fully aware of Hinton's other marriage and, in her testimony at Hinton's trial, she claimed that they had married 'to give a colour of legitimacy' to their children; eight months after the marriage, Maude gave birth to twins.[12] Within three years the pressure of maintaining two households became too much, and Hinton confessed to his first wife and then to a judge. He was tried sentenced to three days in the Pentonville prison in October 1886. The trial was evidently 'managed' by the Hinton family and their connections: the prosecuting solicitor was an old school friend, and both Benjamin Jowett (Master of Balliol College) and Edward Thring (Headmaster of Uppingham) provided character references on Hinton's behalf.[13] At this point Hinton's cultural capital seems to have run dry; unable to find work in Britain after his conviction, he and his first wife emigrated to Japan in 1887, and later

---

9 Henderson, 'Four-Dimensional Space'.
10 For the sake of clarity, I will refer to James Hinton by his full name; when I write 'Hinton', it is in reference to Charles Howard Hinton only.
11 Havelock Ellis notes that this was a topic of conversation over a dinner at the Savile Club with Karl Pearson and Horatio Bryan Donkin; Ellis concluded that the 'syphilitic character' of James Hinton's final illness 'seems to me very doubtful'. See Ellis's journal entry for 2 February 1886, in the Havelock Ellis Papers MSS, ADD 70528, in the British Library.
12 See 'Police', in *The Times* (London), 16 October 1886.
13 See Blacklock's notes on his work in progress, *The Fairyland of Geometry*, particularly his 12 June 2009 entry, 'Rucker on Boole-Stott/Hinton's Bigamy'.

settled in the United States around 1892. There no record of what happened to Maude, although one of Hinton's biographers speculates that Olive Schreiner may have helped her secure passage to South Africa or possibly Australia.[14]

The scandal of the younger Hinton's bigamy conviction guaranteed the association of his unorthodox geometrical theories with his father's theory of 'sexual altruism': 'What a deadly theory that Hinton theory is, like a upas tree blighting all it comes in contact with', Olive Schreiner wrote to Havelock Ellis in 1886, reflecting on the trial, which she attended.[15] After the scandal of the bigamy trial died down, Hinton – along with his father – faded into obscurity in Britain. Karl Pearson, whose work on 'ether squirts' required a theory of the fourth dimension strikingly similar to Hinton's, detested James Hinton and therefore discounted the younger Hinton's work: although Pearson frequently discussed both Hintons in his correspondence during the 1880s, he never in print mentioned Charles Howard Hinton or his work.[16] Even though Hinton had a well-respected American ally in William James, after settling in the United States in 1892 he maintained a low profile, declining James's invitation to give a series of lectures at Harvard.

The primary value of Hinton's work has always been its literary and philosophical content and influence rather than its scientific authority. It is certain that significant late nineteenth-century writers and thinkers such as Wells, William James, Schreiner, Karl Pearson and W. E. B. Du Bois read Hinton. Others, including Henry James, Joseph Conrad and Ford Madox Ford, were familiar with his ideas. Hinton's fourth dimension appealed to scientists, spiritualists and artists, and – particularly at the end of the nineteenth century – the interests of these different groups often overlapped. While not part of Einstein's relativity theory, Hinton's fourth dimension participates in the intellectual trend that Christopher Herbert has identified as 'Victorian Relativity', one which laid the groundwork for the modernist movements of the twentieth century. My project of exploring the literary dimensions of Hinton's fourth dimension is conceived 'in defiance of the founding myth of modernism as a sweeping rejection of Victorian values'.[17] While Hinton and his contemporaries often thought and wrote in reaction to their predecessors, their vocabulary is necessarily informed by the intellectual and moral values of their Victorian and Romantic parents and grandparents. Likewise, the modernists who followed these late Victorians, particularly those who self-consciously rebelled against the values of the previous generation (for example, D. H. Lawrence and Virginia Woolf), at their most experimental and thus reactionary moments were steeped in the language and imagery of the writers I explore in this study. My work here is predicated on the need to 'break down the invidious segregation from one another of different fields of thought', not only between what art and literary critics frequently differentiate as Victorian and modernist periods, but between the arts and sciences.[18] Most specifically, by positioning

---

14 Ibid.
15 Quoted in Grosskurth, *Havelock Ellis*, 102. Taken from the Yaffa Claire Draznin Collection of the Olive Schreiner and Havelock Ellis Papers at the Ransom Center, University of Texas.
16 See Porter, *Karl Pearson*, particularly chapters 6–7.
17 Herbert, *Victorian Relativity*, 29.
18 Ibid., 32.

Hinton's fourth dimension as distinctly *literary*, I am foregrounding the importance of the relationship between mathematical and literary imagination.

Before Einstein: The *Literary* Fourth Dimension

In their introduction to a special issue on the topic of 'Mathematics and Imagination' in the journal *Configurations*, Arielle Saiber and Henry S. Turner cite mathematician Keith Devlin:

> Is there a link between doing mathematics and reading a novel?' Devlin asks. 'Very possibly,' he answers. Imagining a conversation between two invented characters or the intricate imagery of a poem arguably requires a similar kind of mental process as imaging 'the square root of minus fifteen'.[19]

This is an intriguing claim, one that is made explicit in Hinton's theory of the fourth dimension. Hinton recognized the literary nature of his attempts to imagine the fourth dimension when he chose the title *Scientific Romances* for his writings on the subject. According to Hinton and other hyperspace philosophers, the spatial fourth dimension can only be represented in our space as a series of three-dimensional 'slices'. To sustain a representation of these slices in the imagination and fuse them into a whole requires a heroic act of attention very much like the one required of the literary artist in world building, whether that world is the outer social and natural one described by so much nineteenth-century realism or the inner mental world of the modernist individual.

Before advancing further into my discussion of the fourth dimension, it is perhaps necessary to lay one fundamental question to rest, to acknowledge the ambiguity that lies at the heart of the fourth dimension. Does 'the fourth dimension' refer to an actual, 'real' space, or is it an epistemological tool that allows us to better articulate and thus manipulate our environment? To put it simply, when writers refer to 'the fourth dimension', do they refer to something 'out there', something that already exists and is simply waiting for our acknowledgement of it (like X-rays), or do they refer simply to a 'useful fiction', a concept created to provide another way of thinking and talking about physical and psychological sensations? The answer, for Hinton at least, was that such a question is irrelevant: the fourth dimension is both. 'Space' he claimed, 'is the instrument of the mind'.[20] By this he meant literally that space is *the* instrument of the mind; it is a priori, the means by which the mind *thinks* everything else. By adding another dimension to space, we can thus access another dimension of mind. The 'signals which the nerves deliver' to the brain are no more (and probably less, according to Hinton) 'like the phenomena of the outer world' than the shifting bands of colour and black lines of the spectroscope that the astronomer uses to read 'the signal of the skies'.[21] Hinton believed there was a disconnect between the outside world and our mental representation of it.

---

19 Saiber and Turner, 'Mathematics and the Imagination', 1.
20 Hinton, *A New Era*, 2.
21 Hinton, *The Fourth Dimension*, 258.

Thus – in his hyperspace philosophy – for all we know there is a higher, Platonic realm of space out there waiting to be discovered; however, it is only accessible through tuning the 'instrument of the mind'.

Hinton's fourth dimension, like his *Scientific Romances*, functions as fiction. It is through the act of 'reading' that we both create and perceive it. When asked, 'If there are four dimensions, then there may be five and six, and so on up to any number?', Hinton replied that yes, of course, 'when we look quietly at space, she shows us at once that she has infinite dimensions'.[22] However,

> to measure, we must begin somewhere, but in space there is no 'somewhere' marked out for us to begin at. This measuring is something, after all, foreign to space, introduced by us for our convenience. And as to dimensions, in order to enumerate and realize the different dimensions, we must fix on a particular line to begin with, and then draw other lines at right angles to this one. [...] If we take any particular line, we do something arbitrary, of our own will and decision, not given to us naturally by space.[23]

It is the aesthetic will that, by focusing on the fourth dimension, engenders it. Likewise, in his preface to his first novel, Henry James observed that 'really, universally, relations stop nowhere, and the exquisite problem of the artist is eternally but to draw, by a geometry of his own, the circle within which they shall happily *appear* to do so'.[24] Hinton and Henry James (as well as his brother, the psychologist, William) came to the conclusion that the subject, by choosing to fixate on a particular object, in fact *creates* it. The mental process by which one imagines either a fourth spatial dimension or a character's sphere of lived relations is one and the same.

Hinton's project is intertwined with another late-Victorian discourse of relativity, the philosophical school known as pragmatism. Fellow Balliol alumnus and British pragmatist F. C. S. Schiller noted, 'Pragmatism may be taken to point to [...] the plasticity and incompleteness of reality'.[25] William James was a pragmatist, as was his brother, Henry, who observed, after reading his brother's *Pragmatism: A New Name for Some Old Ways of Thinking* (1907), that 'I was lost in the wonder of the extent to which all my life I have [...] unconsciously pragmatised'.[26] For all of these thinkers, the observer is 'the measure of his experience, and so [is] an ineradicable factor in any world he experiences'.[27] Thus, to question whether or not the fourth dimension is 'real' would be to ask Henry James if his art of fiction is 'real': for the pragmatist, the questions are: Are these models relevant to me? Do they allow me to see options in the world previously undetected by me, and are these options – to borrow William James's terminology – 'live' for me?[28] Enacting a functional shift of the word, we might ask, does this novel – or this character – *live*?

---

22 Hinton, 'Many Dimensions', *Scientific Romances*, 39.
23 Ibid.
24 H. James, *The Art of the Novel*, 5, original emphasis.
25 Schiller, *Studies*, 19.
26 H. James to W. James, 17 October 1907, in *Henry James: Letters, 1895–1916*, 466.
27 Schiller, *Studies*, 13.
28 W. James, *Writings, 1878–1899*, 458. See also Carnap, 'Empiricism, Semantics, and Ontology'.

For Henry James, art lives to the extent that it is free.²⁹ A Jamesian character is likewise most 'alive' when he or she is, like Fleda (whom James describes as 'a free spirit'), able to engage with the surrounding social and physical environs (to take a 'contributive and participant view') while aesthetically and morally transcending them.³⁰ Such 'rounded' characters possess freedom of consciousness in E. M. Forster's analysis, as opposed to the 'flat', 'two-dimensional people' who remain circumscribed by their perceptual limitations.³¹ As Mark McGurl observes, Forster's 'terminology is directly descended from the late nineteenth-century preoccupation with dimensionality'; this preoccupation, at its core, was with the possibility that – just as there is an intellectual and aesthetic difference of degree between 'flat' and 'round' characters within a novel – there might exist different dimensions of being between humans outside the novel.³² The sinister possibility that we too are being 'read' by a higher-dimensional consciousness is also implied here.

'Ambulatory Relations'

My organization of this book is informed by William James's pragmatic 'ambulatory' methodology, which Henry embodied in aesthetic practice. William James argued, 'of the relation [to the world] called "knowing," which may connect an idea with a reality':

> My own account of this relation is ambulatory through and through. I say that we know an object by means of an idea, whenever we ambulate towards the object under the impulse which the idea communicates. If we believe in so-called 'sensible' realities, the idea may not only send us towards its object, but may put the latter into our very hand, make it our immediate sensation. [...] The idea is thus, when functionally considered, an instrument for enabling us the better to *have to do* with the object and to act about it. But it and the object are both of them bits of the general sheet and tissue of reality at large; and when we say that the idea leads us towards the object, that only means that it carries us forward through intervening tracts of that reality into the object's closer neighbourhood [...]. My thesis is that the knowing here is *made* by the ambulation through the intervening experiences.³³

Building on the work of Richard A. Hocks, I interpret this methodology of 'ambulatory relations' as a constant reassessment of familiar texts and ideas in light of fresh evidence.³⁴ Throughout this book I demonstrate how Hinton employs a similar strategy in his hyperspace philosophy, asking his readers to ambulate through his various texts while re-reading and re-considering previous ideas in light of increased higher-dimensional knowledge. My choice of this approach is informed by the dual need to introduce Hinton and his ideas while demonstrating the literary quality and aesthetic impact of his fourth dimension. The book is divided into two parts; the first introduces the theory of

---

29 Matthiessen, ed., *The James Family*, 357.
30 H. James, *The Art of the Novel*, 131 and 130.
31 Forster, *Aspects of the Novel*, 95.
32 McGurl, *The Novel Art*, 64.
33 W. James, *Writings, 1902–1910*, 898–99, original emphasis.
34 Hocks, *Henry James and Pragmatistic Thought*, 38–47.

the fourth dimension, Hinton and his hyperspace philosophy. The *Scientific Romances* of Hinton are given careful attention here as well. In the second part of the book, I traverse the writings of William James, H. G. Wells and Henry James: the work of these writers is read 'through' the four-dimensional aesthetic of Hinton's hyperspace philosophy. Each chapter builds upon and revisits the previous ones, mimicking the ambulatory process by which Hinton introduced his readers to the fourth dimension.

Part I: Reading the Fourth Dimension

The first chapter establishes the roots of Hinton's hyperspace philosophy, tracing the evolution of the idea of the fourth dimension from the abstract language of analytical geometry to descriptive geometry, and into contemporary debates over the 'new', non-Euclidean geometries. These debates and the anxieties underpinning them were surprisingly productive, generating the fantasy spaces of Lewis Carroll's Wonderland and Looking-Glass Land, as well as the fourth dimension of hyperspace philosophy. Through examining contemporary discussions of space, I trace the movement of 'flatland narratives' out of scientific and philosophical journals such as *Nature* and *Mind*, and into popular literary discourse. This initial chapter also establishes Hinton's own hyperspace philosophy in relation to the culture of Oxford in the 1870s, from his exposure to Thomas Hill Green's lectures on Kant at Balliol College, to Hinton's involvement with Ruskin as a key player on the Hinksey road project. Hinton was at Oxford from 1871 to 1876, during the period when both Aesthetes and Idealists were discussing theories of perception as an act of creation.

Drawing on Gillian Beer's important work identifying the key methodological questions facing scholars in the field of literature and science, Alice Jenkins argues that a sense of the diffuseness of the reception of ideas in nineteenth-century culture is best expressed not in the 'traditional "history of ideas" model of dissemination', but that 'a more accurate sense of the movement of ideas from context to context within the period would emphasize the accidental, the partial, and the metaphorical'.[35] William James's ambulatory methodology is apt here, and in my discussion in Chapter One (and throughout this book) I pay careful attention to the unintentional and felicitous movement of the concept of the fourth dimension from its mathematical origins through the discourses of physics to idealist and socialist philosophies, as well as aesthetics.

In Chapter Two, I turn to the first series of Hinton's *Scientific Romances*, a series of pamphlets published between 1884 and 1886, which were bound and sold as a complete volume from 1886 onward. The texts that make up the first series include, among others, a philosophical meditation, 'What is the Fourth Dimension?', an allegorical tale, 'The Persian King', and the first of Hinton's cube exercise manuals, 'Casting Out the Self'. These texts function together in forming an unstable unity; each individual 'romance' plays off the others, and in reading these texts, one is pushed to test Hinton's hypothesis that the fourth dimension can be perceived from a three-dimensional perspective only as

---

35 Jenkins, *Space*, 141–42. See also Beer, *Open Fields*, 173–95.

a series of 'slices'. In each of these texts, Hinton sought to describe the fourth dimension to his readers and to guide them toward forming a representation, for themselves, of four-dimensional existence.

This idea of ambulation through 'slices' of experience is what W. D. Howells had in mind when he wrote that his piecemeal critical appreciation of Dante's *Divine Comedy* was superior because 'we see nothing whole, neither of life nor art. We are so made, in soul, and in sense, that we can only deal with parts, with points, with degrees':

> I am very glad that I did not then lose any fact of the majesty, and beauty, and pathos of the great certain measures for the sake of that fourth dimension of the poem which is not yet made palpable or visible.[36]

In opposition to Howells, Hinton intended the piecemeal process of perceiving the fourth dimension to be a means rather than an end; it is this 'fourth dimension' of life and art that Hinton wanted to eventually reveal through his writings. Paradoxically, however, this is a never-ending procedure, and – from a critical perspective – it is Hinton's focus on aesthetic process that is the most interesting aspect of his hyperspace philosophy.

In his attempts to create new rules for seeing, Hinton expected his readers to undergo a process that Wolfgang Iser describes in his theory of aesthetic response:

> In literature, where the reader is constantly feeding back reactions as he [*sic*] obtains new information, there is [...] a continual process of realization, and so reading itself "happens" like an event, in the sense that what we read takes on the character of an open-ended situation, at one and the same time concrete and yet fluid. The concreteness arises out of each new attitude we are forced to adopt toward the text, and the fluidity out of the fact that each new attitude bears the seeds of its own modification. Reading, then, is experienced as something which is happening – and happening is the hallmark of reality.[37]

The juxtaposition of genre and style in Hinton's collection of essays, meditations, tales and cube exercises creates a similar feedback loop, thus enabling the reader to construct the 'reality' of the fourth dimension. In Chapter Two, I argue that the effect of Hinton's individual texts, once collected together in the *Scientific Romances*, is to engender an overtly 'open-ended situation' in which the reality of the fourth dimension is allowed to develop within the reader's mind through a process of analogical construction, deconstruction and correction.

Chapter Three provides another ambulation through Hinton's literary fourth dimension, this time by exploring his second series of *Scientific Romances*. While all of the texts in the first series were composed and published before Hinton's bigamy conviction, the second series consists of texts composed in Britain, Japan and the United States. My main focus in this chapter is on two novellas from this series, *Stella* and *An Unfinished Communication*, which were originally published together in 1895. These texts mark an

---

36 Howells, *My Literary Passions*, 202.
37 Iser, *The Act of Reading*, 68.

inward turn for Hinton; while in the first series he was primarily concerned with conceiving and perceiving the fourth dimension, in the second series he explicitly began to explore the social and psychological implications of his hyperspace philosophy. I read *Stella* in particular alongside contemporary debates within radical fin-de-siècle Britain, with particular focus on the writings of Havelock Ellis, Edward Carpenter and Friedrich Nietzsche. The sexual and socialist politics of James Hinton and his circle resonate throughout *Stella*, which tells the story of a young woman who is made invisible by an older man as a socialist experiment in overcoming 'self-regarding impulses'.[38]

In Chapter Three I also examine *An Unfinished Communication*, the other novella of the second series, a text that is arguably Hinton's highest literary achievement. Offering one of the earliest English-language engagements with Nietzsche's ideas, Hinton dramatized his protagonist's perception of the fourth dimension – which occurs through his experience of external recurrence – as the discovery of his own transcendental will-to-power. In *Thus Spake Zarathustra*, Nietzsche performed his 'thought-experiment' of eternal recurrence in order to see, as Matthew Rampley argues, 'how "incorporation" of the idea of Eternal Recurrence would *change* and *alter* human thinking and practices'.[39] In Hinton's version of this thought-experiment, his protagonist is able to access a 'wider view' of his life, depicted as a kind of suprahistorical perspective of all recurrences of that life. Time is reduced to space here, and through this spatio-temporal view from the fourth dimension, Hinton's protagonist is able to 'unlearn' his nostalgia for the past and take an active role in shaping his future.

Part II: Reading *Through* the Fourth Dimension

Hinton, Wells, and the James brothers were concerned with the evolution of human consciousness, and all four expressed what they perceived to be the highest form of consciousness as a kind of will-to-create. In Part II, I read the work of these three writers through the lens of the four-dimensional literary aesthetic established in Part I. I begin Chapter Four by examining the correspondence between Hinton and William James: both the actual remaining letters from Hinton to William James, and 'the accidental, the partial and the metaphorical' correspondences between Hinton's fourth dimension and James's pragmatist model of consciousness as movement. I identify their shared recourse to Gustav Theodor Fechner's 'mother-sea' metaphor of consciousness in their attempts to represent what both men saw as another dimension of being, which occasionally, through heightened, 'supernatural' experiences, irrupted into the consciousness of extraordinary individuals. Fechner's metaphor (also described as a 'wave-scheme' by William James) reverberates throughout the writings of Pater, Woolf and Freud as well.

In Chapter Five I turn to William James's admirer and fellow scientific romancer, H. G. Wells. Here I establish the case for Hinton's influence on Wells but, more importantly, I examine how Wells's *The Invisible Man* (1897) responded to the same social and sexual politics raised in *Stella*. Importantly, Wilhelm Röntgen published his discovery of

---

38 Hinton, *Stella, Scientific Romances*, 48.
39 Rampley, *Nietzsche, Aesthetics and Modernity*, 149, original emphasis.

X-rays after *Stella*, but before *The Invisible Man* was written. While (as I demonstrate in Chapter Three), Hinton drew upon recent innovations in moving picture and phonographic technologies to express his idea of the fourth dimension, Wells's text is steeped in the anxieties and excitement raised by the contemporary 'mania' for X-ray photography. The discovery of X-rays lent credence to the claims of Hinton and others that a four-dimensional being, like a three-dimensional human looking down on a world of two dimensions, would be able to see all parts of a solid figure within our three-dimensional space. Four-dimensional 'vision' thus corresponds to 'X-ray vision' and brings with it all the anxieties that come with the prospect of being subjected to a penetrating, panoptic gaze. It is Wells's protagonist's discovery of 'a general principle of pigments and refraction, – a formula, a geometrical expression involving four dimensions', that allows him to render himself invisible.[40] By reading Wells's experiment in invisibility alongside Hinton's *Stella*, I identify what is implicit in both writers: the 'othering' effect of the fourth dimension. Bound up with pathological discourses of 'sexual inversion', feminine 'nature' and evolutionary racism, both Wells and Hinton demonstrated that exposure to the fourth dimension can result in 'inversion' for the three-dimensional subject. In the case of the flatland narratives of Hinton and Wells, this inversion is literal (movement through a higher dimension results in the lower-dimensional being's 'turning over', or flipping, so that its right and left sides are inverted), while in their narratives of invisibility this 'inversion' is subtler.

My readings of Wells's early fictions are drawn together in my exploration of what William J. Scheick has described as Wells's 'four-dimensional' literary aesthetic, of 'splintering' the narrative frame of his fictions.[41] Wells developed this approach in opposition to Henry James's aesthetic, and their debate over this matter (as well as their friendship) came to a messy end with Wells's infamous critique of James in his 1915 novel, *Boon*. I conclude Chapter Five by turning to *Boon*, to re-examine Wells's quarrel with Henry James over the art of fiction in light of Wells's experimentation in the literary fourth dimension.

Chapter Six, the final chapter, is an exploration of how hyperspace philosophy, and 'the late-nineteenth-century preoccupation with dimensionality', resonates within James's late style. Of all the writers I examine throughout *Before Einstein*, Henry James is the only one not directly acquainted with Hinton, either personally or through Hinton's work. James was, however, familiar with the concept of the fourth dimension. Reading Henry James through the lens of the fourth dimension allows me to test to what extent Hinton's hyperspace philosophy (and the methodology of ambulatory relations) can function as a critical apparatus. James's later fiction is, as Hazel Hutchison and others have noted, concerned with 'the role that language plays in constructing the fictional world'.[42] For Henry James, the rhetoric of space and ambulation was useful in constructing his fictional worlds; thresholds became important in James's later fiction, and he often staged key 'realization' moments for his central characters as the penetration of elaborate

---

40 Wells, *The Invisible Man*, 89.
41 Scheick, *The Splintering Frame*.
42 Hutchison, *Seeing and Believing*, 3.

perceptual frames. For example, noting Maggie's position on the balcony looking in on the scene between her husband, father and her best friend/stepmother at the end of *The Golden Bowl* (1904), McGurl observes that Maggie's physical position 'is a figure of her access to the "space apart" that is consciousness'.[43] As she watches the scene indoors, Maggie comes to a greater understanding of the sexual intrigues to which she has been blind throughout the novel; she, like Fleda, becomes the 'one who knew the most'. In the knowledge-as-power struggle that results after her stepmother, Charlotte, joins her on the balcony, Maggie triumphs, becoming the author of her own scene. She is able to discern, and thus manipulate, the situation more subtly than could Charlotte. Significantly, for my four-dimensional reading of James, McGurl asks:

> But where, after all, is the mind that makes distinction? Where is the scene of metaphysical vision? Can it actually be put into a book? Or does it hover at the surface of the text, looking down upon it as James looks down upon the characters he has created? That is one of the issues raised by James's representation of Maggie and Charlotte standing on the outside looking in, for though both are physical 'outsiders' in this scene, only one of them is also an 'insider' to the knowledge that the room, like the novel itself, contains.[44]

It is this mind, this metaphysical (perhaps 'metatextual') presence hovering over the space of the text that Fleda perceives on the threshold of the house at Ricks. Spatial relations in the Jamesian novel tend to denote a hierarchy of knowledge, much as they do in the dimensional analogy of flatland narratives. While Fleda can sense 'a kind of fourth dimension' and therefore reveals the 'comparative stupidity' of the other characters in *The Spoils of Poynton*, the author necessarily possesses an even 'higher' intelligence.

In other texts from the turn of the century, James utilized spatial rhetoric to explore the fiction-creating capability of consciousness: in 'The Great Good Place' (1900), he removed his central character, George Dane, from 'representational' space of his everyday world in the south of England and located him in a quasi-supernatural 'place'. It is 'much nearer than one ever suspected [...] nearer everything – nearer every one', one character explains, and yet it lies outside all known space and time.[45] By escaping to this place for a period of time, Dane – himself an author – is able to recover his creative agency. We can read the space of this 'Great Good Place' as four-dimensional, the 'site' of aesthetic creation itself. Such is the kind of fourth dimension D. H. Lawrence had in mind in his claim that

> when Van Gogh paints sunflowers, he reveals, or achieves, the vivid relation between himself, as man, and the sunflower as sunflower [...]. You cannot weigh nor measure nor even describe the vision on the canvas. It exists, to tell the truth, only in the much-debated fourth dimension.[46]

---

43 McGurl, *The Novel Art*, 54.
44 Ibid., 56.
45 H. James, *The Novels and Tales*, 16: 238.
46 Lawrence, *Study of Thomas Hardy*, 171.

In Lawrence's interpretation, the fourth dimension is the 'space' of pure relations between artist and object. This extra-representational locale is coded as superior to the three-dimensional material world, just as our world is to a shadowy realm of two dimensions. Thus it is appropriate that I conclude Chapter Six by exploring the house in 'The Jolly Corner' (1908) as an uncanny house of fiction, a borderland 'space' of relations between the author and his creation.

The Literary Fourth Dimension

'We have yet fully to explore the cultural life of the imaginary, the hypothetical, and the abstract spaces in which no nineteenth-century person walked, but with and through which they thought', Jenkins observes.[47] I describe the fourth dimension under discussion in the present book as literary, not just because it is one of the imaginary spaces to which Jenkins refers, but because it is a space of shifting meaning, a metaphor that travels across discourses and through which we can mentally ambulate. This fictional space serves as fertile ground for the creative thinker; it is equally appealing for its radical equalizing capabilities and its potential as a space of elite aesthetic sensibility.

Before moving onward to consider the first movement of the fourth dimension from the abstract language of analytical geometry to concrete concept, I conclude this introduction by demonstrating briefly how reading through the fourth dimension can reveal fresh interpretations of familiar writers. The most obvious example is H. G. Wells: while several critics have at least briefly mentioned his discussion of the fourth dimension in *The Time Machine* (1895), the often overlooked – but much more fruitful – candidate for analysis of Wells's early romances is his other 1895 book, *The Wonderful Visit*. This text tends to be neglected, perhaps because it is atypical of Wells's proto-science fiction. However, *The Wonderful Visit* is fascinating in part because it is such an anomaly in Wells's œuvre. Here we can observe his flirtation with late nineteenth-century Aesthetic and Decadent movements and his unsympathetic presentation of their critics. Most importantly for my study, this text links the heightened sense of beauty of the aesthete with a four-dimensional, 'higher' consciousness.

The premise of *The Wonderful Visit* is simple but bizarre: an angel from the fourth dimension is shot down to earth. Wells claimed that the idea for this story 'was obtained from Ruskin's assertion that if an angel were to appear on earth someone would be sure to shoot it'.[48] Indeed, through his use of stereotypes alone, Wells appeared to be in sympathy with Ruskin's protest against the prosaic nature of late-Victorian culture: the shooter in his novel is the vicar of a small parish in the south of England, who is an amateur ornithologist. The inability of the provincial English villagers to recognize the higher spirit of truth and beauty is a theme that informs the entire novel, as the 'fallen' angel is misunderstood and even persecuted by the local villagers. What is most striking

---

47 Jenkins, *Space*, 234.
48 Quoted in Raknem, *H. G. Wells and His Critics*, 417.

about this angel is, however, the fact that he is not of heavenly origin. The narrator interrupts the story to explain:

> Let us be plain. The Angel of this story is the Angel of Art, not the Angel that one must be irreverent to touch – neither the Angel of religious feeling nor the Angel of popular belief. [...] This Angel the Vicar shot is, we say, no such angel at all, but the Angel of Italian art, polychromatic and gay. He comes from the land of beautiful dreams and not from any holier place. At best he is a popish creature.[49]

The angel is very much a creature of 1890s aesthetic sensibility and would not appear out of place in Wilde's circle: he is 'a youth with an extremely beautiful face, clad in a robe of saffron and with iridescent wings, across whose pinions great waves of colour, flushes of purple and crimson, golden green and intense blue, pursued one another'.[50] His mannerisms are strikingly fey: he frequently laughs with amusement at the novelty of life on earth, finding it 'delightfully grotesque' at first. Because of his delicate appearance he creates a brief scandal for the vicar when he mistaken for a young woman by the curate's wife and her visitors. He is also a musical genius able to play the violin by ear, and the self-appointed intellectuals of the village – who refuse to believe the vicar's claims that the angel is in fact an angel – interpret his strange appearance and bohemian behaviour as indicative of his aesthetic genius, as well as his mental and moral degeneracy. Max Nordau, who famously pathologized Wilde in *Degeneration*, is cited with authority by the village doctor when he is attempting to diagnose the cause of the angel's strange 'wing-like appendages' and his inability to understand polite social codes. 'For a moment', John Batchelor observes, Wells 'seem[ed] to join Wilde, Beardsley, Max Beerbohm and the rest in teasing the bourgeoisie from the standpoint of the aesthete'.[51] Indeed, it seems likely that Wells's depiction of the villagers' perception of the threatening and offensive otherness of the angel, along with their persistent hounding of him and his eventual demise, were influenced by the fact that he was writing this novel either during or shortly after the Wilde trials.

Strikingly, Wells codes the angel's otherness as resulting from his higher-dimensional nature. The vicar, who is also an amateur geometer, comments that hearing of the angel's inexplicable movement from his own world to the vicar's 'almost makes one think there may be [...] Four Dimensions after all'.[52] The angel has somehow accidentally accessed this four-dimensional space and become trapped in the vicar's world. While here Wells provides an explicit link to contemporary hyperspace discourse, what is more relevant to this discussion is his description of the angel's world, the 'land of beautiful dreams', where

> there is nothing but Beauty there – all the beauty in our art [on earth] is but feeble rendering of faint glimpses of that wonderful world, and our composers, our original composers, are those who hear, however faintly, the dust of melody that drives before its winds.[53]

---

49 Wells, *The Wonderful Visit*, 35, 37.
50 Ibid., 15.
51 Batchelor, *H. G. Wells*, 8.
52 Wells, *The Wonderful Visit*, 26.
53 Ibid., 28–29.

The idea of earthly beauty being only a 'faint glimpse' of the beauty of the angel's world echoes Hinton's writings in which he frequently employed Plato's allegory of the cave to illustrate the relationship between the 'higher reality' of the fourth dimension and our own world. Wells's angel comes from a similarly Platonic realm of art, of which the arts in the vicar's world are only feeble shadows. This association of four dimensionality with a higher aesthetic sensibility is not simply an obscure connection made by the young fantasist Wells: one week after Wilde's conviction, Ernest Newman wrote in the *Free Review* of Wilde's statement that 'All Art is immoral':

> If a thinker says 'Art is immoral', the new synthesis puzzles [the majority], and they either call it a paradox, or say the writer is immoral. In reality, he is just doing what they cannot do; he can see round corners and the other side of things. Nay, he can do more than this; he can give to ordinary things a quality that they have not, and place them in worlds that never existed. We ordinary beings can see objects in three dimensions only; a good paradox is a view in the fourth dimension.[54]

Newman's statement nicely brings together the ways in which four-dimensional theory, aesthetics, 'otherness' and, later, X-ray vision, mutually reinforced each other.

In these elite (and sometimes elitist) discourses, the extra-representational space of the fourth dimension is coded as aesthetically and morally superior to the three-dimensional material world, just as our world is compared to the shadowy two-dimensional realms of the flatland narratives. The extra-representational nature of the fourth dimension is at the root of its greatest paradox and its greatest interest for literary scholars. In his reading of Abbott's *Flatland* and James's *The Princess Casamassima* (1888), McGurl demonstrates one way we might consider this extra-representational space in relation to fiction:

> The inhabitants of Flatland exist as 'characters' in two senses of that term, both as represented beings and as conventional symbols, somewhat as though the type beneath our eyes has detached itself from the pulp upon which it is pressed and come to life. It is a bizarre form of life, lived laterally, confined to the two-dimensional plane of the page.[55]

The dimensional analogy of the flatland narrative was consistently deployed by hyperspace philosophers to represent the extra-representational fourth dimension. This analogy, as implied in McGurl's statement, is particularly well adapted to arts that are confined to a two-dimensional surface, such as painting or writing. If what is depicted on the two-dimensional surface of the page or the canvas is supposed to be representational of the three-dimensional world, then it is not difficult to imagine the possibility of a higher dimension outside or above our space. Just as many cubist painters were trying to represent a four-dimensional perspective in their work during the early years of the twentieth century, some writers attempted to represent the experience of a three-dimensional character encountering a four-dimensional space or presence in the 1880s and 1890s.

---

54 Newman, 'Oscar Wilde', 233. Born William Roberts, Newman adopted his pen name in the 1890s; his choice is indicative of the popularity of 'new' as a descriptive during this time.
55 McGurl, *The Novel Art*, 57.

This moment of contact served as a metaphor for the encounter between the creative will and world it has created. Often, Henry James coded this confluence as 'ghostly', though not necessarily supernatural. For the Jamesian central character, these ghostly encounters serve, Timothy Lustig writes, to represent 'a particularly intense adventure of consciousness, an access of liberate and disencumbered experience, [that] one could argue' brings the seer 'extremely close to James himself'.[56]

While his brother was dramatizing such encounters in fiction, William James was exploring 'ghostly' encounters as a psychologist, philosopher and president of the British Society for Psychical Research. In an 1895 lecture, James explained that 'so far as man stands for anything, and is productive or originative at all, his entire vital function may be said to have to deal with maybes': in cases in which definitive support for or against a hypothesis is lacking, human beings have a right to act on whichever alternative is most conducive to their survival. By so believing, he told his audience, 'you make one or the other of two possible universes true by your trust or mistrust – both universes having been only *maybes*, in this particular, before you contributed your act'.[57] A year later, William James described this as 'the will to believe' in a lecture of the same title. James understood that the will to believe was a creative one; it was only by looking for a world that one was likely to find it. Even beliefs that were acknowledged to be fictions could be 'useful fictions'. It is only after reading Hinton's hyperspace philosophy for its literary as well as conceptual content that we can begin to understand how the idea of a spatial fourth dimension detached from its origins in nineteenth-century non-Euclidean geometry and became just such a 'useful fiction' for writers and artists at the turn of the twentieth century.

---

56 Lustig, *Henry James and the Ghostly*, 63.
57 W. James, *Writings, 1878–1899*, 500–501, original emphasis.

Part I

READING THE FOURTH DIMENSION

Chapter One

IMAGINING 'SOMETHING PERFECTLY NEW': PROBLEMS OF LANGUAGE, CONCEPTION AND PERCEPTION

Ezra Pound's call to his contemporaries to 'make it new', although suggesting avant-garde intent, was actually part of a concentrated interest in 'the new' in Anglo-American culture and is traceable as far back as at least the 1880s.[1] As Holbrook Jackson observed in 1913, the popularity of the adjective new grew during the fin de siècle.[2] Writing of the New Realism in 1897, H. D. Traill claimed that 'not to be new is, in these days, to be nothing'.[3] Other notable examples of the vogue of the new are the New Spirit, the New Drama of Ibsen and, of course, the New Woman. It is not surprising then that a 'new geometry' would appeal to this generation of writers and thinkers.[4] It is in this context that we should consider Charles Howard Hinton's hyperspace philosophy, which was first fully expressed in *A New Era of Thought* (1888). In this book he promised to 'bring forward a complete system of four-dimensional thought – mechanics, science, and art'.[5] While Hinton did not live to complete this system, his belief in the applicability of 'four-dimensional thought' across multiple discourses was appropriate: the history of the concept of the spatial fourth dimension is a history of movement. It is also part of the shared history of modernism.

The rise of non-Euclidean geometry in the second half of the nineteenth century served to emphasize the contingency of even mathematical knowledge, pushing debates about the relativity of knowledge to the forefront in a way that must have been particularly distressing for conservative thinkers. Euclid's axioms, which had remained largely uncontested for nearly two thousand years, were no longer sacrosanct. 'The argument concerning the relativity of knowledge is absolutely necessary to the emergence of modernism,' Gillian Beer correctly explains, finding 'the cognate confusion between method and findings' in late Victorian mathematics and physics particularly suited for uncovering

---

1 Pound first used this phrase in *The Cantos*. However, he borrowed this slogan from Cheng Tang, the founder of the Shang dynasty. Thus, while this phrase is associated with an earlier 'break' with the past, it is also a call for renewal, or recurrence with variation. See Sun, 'Pound's Quest for Confucian Ideals', 96–119.
2 Jackson, *The Eighteen Nineties*, 23.
3 Traill, *The New Fiction*, 1.
4 Non-Euclidean geometry is described as the 'new geometry' as early as 1865; German mathematician Julius Plücker lectured 'On the New Geometry of Space' to the London Royal Society in February of that year. However, the term was not used frequently until the 1890s.
5 Hinton, *A New Era*, 86.

connections with 'proto-modernist texts'.[6] The first part of the present chapter traces the movement of the concept of the fourth dimension from its origins in analytical geometry to its leap to narrativization via the dimensional analogy; in the second part I consider Hinton's particular interpretation of the fourth dimension in light of his early intellectual influences, including James Hinton, Ruskin and Kant.

The New Geometries

In *The Fourth Dimension and Non-Euclidean Geometry in Modern Art*, Linda Dalrymple Henderson connects the shift from high Victorian realism to more abstract forms of art, generally described as modernist, to a similar shift in late nineteenth-century geometry.[7] However, more was at stake in the challenge the new geometries presented to Euclid than aesthetics or mathematics. Alice Jenkins has uncovered the hidden dimension of class politics in Euclidean geometry, noting how in the early nineteenth century 'mathematics held an immensely privileged status in the European concept of education, and at the root of its status lay the classical study of geometry'.[8] Knowledge of classical languages and higher mathematics was the hallmark of the Oxbridge-educated male, and debates around the utility of Euclidean geometry in education and the applied sciences were necessarily underpinned by questions of class. At the polar ends of this debate were the classicists, who argued that the study of geometry was fundamental for developing the faculty of reason, and those who argued that the importance of higher mathematics in education and culture was greatly overemphasized by the privileged classes. 'In between these two positions', Jenkins observes,

> were more moderate views which broadly supported the study of geometry but sought to divest it of its aura of privilege and inaccessibility by teaching in such a way as to emphasize practical rather than abstract reasoning (and thus, to the adherents of the Euclidean method, denuding it of most of its benefit to the learner).[9]

Educational reform debates continued into the second half of the century, and it was clear which side was winning when T. H. Huxley began to emphasize the importance of early education in the physical sciences over abstract mathematics. In his address to the Liverpool Philomathic Society in 1868 (later published in *Macmillan's Magazine*), Huxley lamented the lack of practical scientific training in primary and secondary education. According to Huxley, the wealth and health of the nation depend on early scientific training, and this training must be practical, not abstract, 'bringing [...] the mind directly into contact with fact, and practising the intellect in the completest form of induction; that is to say, in drawing conclusions from particular facts made known by immediate observation of nature'.[10] The study of mathematics would not offer the same kind of

---

6 Beer, *Open Fields*, 303.
7 Henderson, *The Fourth Dimension*, 98.
8 Jenkins, *Space*, 166.
9 Ibid., 167.
10 Huxley, 'Scientific Education', 182.

discipline: 'mathematical training is almost purely deductive. [...] There is no getting into direct contact with natural fact by this road'.[11]

With the tide turning in favour of practical scientific training, mathematicians such as James Joseph Sylvester sought to defend mathematical training by adapting and subverting Huxley's argument. The classicist Euclideans were losing the battle: in his 1869 address to the Mathematical and Physical Section of the British Association for the Advancement of Science (BAAS), even Sylvester claimed he would like to see 'Euclid honourably shelved or buried [...] out of the schoolboy's reach'.[12] Nevertheless, he directly challenged Huxley's claim that 'mathematical training is almost purely deductive':

> Mathematical analysis is constantly invoking the aid of new principles, new ideas, and new methods, not capable of being defined by any form of words, but springing direct from the inherent powers and activity of the human mind, and from continually renewed introspection of that inner world of thought of which the phenomena are as varied and require as close attention to discern as those of the outer physical world [...]: that it is unceasingly calling forth the faculties of observation and comparison, that one of its principal weapons is induction, that it has frequent recourse to experimental trial and verification, and that it affords a boundless scope for the exercise of the highest efforts of imagination and invention.[13]

The shift in tone is subtle but important: within this plea for the recognition of the value of introspection in scientific education, Sylvester adopts the very terms of Huxley's argument that inductive reasoning is superior to deduction. Its place no longer assured in the highest reaches of intellectual respectability (or the foundations of educational training), mathematics is legitimized here as an analogue to the natural sciences: Sylvester even went so far as to describe Arthur Cayley as 'the central luminary, the Darwin of the English school of mathematicians'.[14]

We should consider Hinton as an inheritor of this shifting debate: although the fourth spatial dimension was accepted by most reputable mathematicians and scientists as purely theoretical, Hinton argued for the discernment of higher space through practical training. His hyperspace philosophy, although dealing with what many would call abstract space, was the product of these attempts to emphasize the practical applications of geometry and confusions arising from the increasingly specialized and abstract nature of mathematical, particularly algebraic, discourse. Sylvester's address demonstrates how the climate was ripe for the confusion of abstract terms with practical applications. After lamenting that even 'authorized' English writers such as William Whewell, G. H. Lewes and Herbert Spencer conflate the terms 'reason' and 'understanding', or 'Vernunft' and

---

11 Ibid.
12 Sylvester, 'A Plea', 2: 261. Sylvester clarifies that 'I have used the word mathematics in the plural; but I think it would be desirable that this form of word should be reserved for the applications of the science, and that we should use mathematic in the singular number to denote the science itself' (262).
13 Sylvester, 'A Plea', 1: 237.
14 Ibid., 238.

'Verstand', Sylvester celebrated the unification of the 'matter and mind' of the various branches of mathematics:

> Time was when all the parts of the subject were dissevered, when algebra, geometry, and arithmetic either lived apart or kept up cold relations [...]; but that is now at an end; they are drawn together and are constantly becoming more and more intimately related and connected by a thousand fresh ties, and we may confidently look forward to a time when they shall form but one body with one soul.[15]

Hinton's fourth dimension arose from the conflation of algebraic terminology and descriptive geometry. For example, in seeking to find the geometric figure corresponding to x^4, Hinton coined the term 'tesseract', indicating a four-dimensional analogue to the cube, or x^3.

When Hinton came of age, non-Euclidean geometry was just reaching popular scientific discourse. Although non-Euclidean geometry was simultaneously and independently 'discovered' by Johannes Bólyai and Nicholai Lobachevskii in the 1820s, it did not enter mainstream mathematics in Britain until 40 years later. At this time in curriculum reform debates, the classicist Euclidean method was under attack. Jonathan Smith observes:

> In a country where a staple of education from the lower forms to the universities was the study of Euclid's *Elements*, the development of different geometries and the contention that space may not be Euclidean and three-dimensional could not help but capture public attention.[16]

Smith's grouping of Euclidean and three-dimensional geometry also illustrates the way the public conflated non-Euclidean geometries with the theory of the fourth dimension. From the 1870s onward, a growing body of specialist and popular literature that addressed the new geometries often combined the concepts of the fourth dimension and *n*-dimensional spaces with non-Euclidean geometry. Although the possibility of *n*-dimensional spaces was only one idea raised within specialist discussions of non-Euclidean geometry, it soon became representative of these new geometries to popular audiences. For many, the concept of *n* dimensions itself was understood as the theory of the fourth dimension of space. While most specialists understood the difference, as K. G. Valente has shown, these mathematicians often unintentionally implied a relationship between non-Euclidean, curved models of space and the fourth dimension. Hermann von Helmholtz, W. K. Clifford and other mathematicians,

> as part of their mission to disseminate radically new geometric epistemologies to a wider audience [...] often asked their readers to contemplate the limited understanding that beings living on the two-dimensional surface of a sphere would have of the curved geometry of their world [...]. This illustrative scenario was meant in part to show how one could understandably mistake our space as Euclidean [...] based on small-scale experiences or observations. It gave rise, however, to a commonly held misconception [...]. Consequently, promoting

---

15 Ibid., and Sylvester, 'A Plea', 2: 262.
16 Smith, *Fact and Feeling*, 180.

non-Euclidean or Riemannian models of space in the 1870s simultaneously, if unintentionally, served to draw attention to the fourth dimension.[17]

In this way, the fourth dimension came to be associated with both non-Euclidean geometries and n-dimensional geometries.

N-dimensional (or sometimes, 'p-dimensional') spaces had more or less than three dimensions and were considered to be purely analytical and abstract by most mathematicians and scientists. The potential for reification of these terms occurred in the shift from the analytical language of algebra to the more descriptive language of geometry. In her study of Victorian geometry, Joan Richards explains this difference: 'Geometrical arguments are clearly more descriptive than analytical [algebraic] ones. To argue that a proof involving circles requires a conception of space is much easier than arguing that an analytical demonstration involving a and b requires an understanding of number.'[18] The concept of the fourth dimension of space grew out of a slippage between these discourses; it was the result of a hypostasization of abstract symbols such as x^4.

The potential for such slippage was present in the writings of Victorian geometers, as Richards shows in an example taken from an 1866 essay by the mathematician George Salmon, 'On Some Points in the Theory of Elimination':

> The question now before us may be stated as the corresponding problem in space of p dimensions. But *we consider it as a purely algebraical question, apart from any geometrical considerations*. We shall however retain a little of the geometrical language, both because we can thus avoid circumlocutions, and also because we can more readily see how to apply to a system of p equations, processes analogous to those which we have employed in a system of three.[19]

In this passage, Salmon was specific that he was not referring to an actual space of p dimensions; rather, he was considering a purely formal problem. For him, the language of descriptive geometry was simply a matter of convenience. However, Richards observes, although Salmon was clear that 'he was just using a figure of speech [...] Cayley was less explicit on this point'.[20] This ambiguity on Cayley's part did not pass unnoticed by other British mathematicians. In his 1869 address to the BAAS cited above, Sylvester actually made the jump from an abstract treatment of n dimensions to a suggestion of the 'reality of transcendental space' of four or more dimensions.[21]

As Richards notes, Sylvester's support for the reality of higher spatial dimensions was 'rather circuitous'.[22] Rather than attempt to illustrate his own conception of four or more dimensions, Sylvester cited Gauss and Cayley as key supporters. Additionally, in a

---

17 Valente, 'Who Will Explain the Explanation?', 130.
18 Richards, *Mathematical Visions*, 39.
19 Salmon, quoted in *Mathematical Visions*, 54, emphasis added. The essay originally appeared in an 1866 issue of the *Quarterly Journal of Pure and Applied Mathematics*. Salmon's choice of the variable p is arbitrary and interchangeable with n.
20 Richards, *Mathematical Visions*, 55.
21 Sylvester, 'A Plea', 1: 238.
22 Richards, *Mathematical Visions*, 56.

footnote he mentioned Clifford in conjunction with speculations about the fourth dimension, suggestively remarking:

> If an Aristotle or Descartes, or Kant assures me that he recognises God in the conscience, I accuse my own blindness if I fail to see him. If Gauss, Cayley, Riemann, Schalfi, Salmon, Clifford, Krönecker, [sic] have an inner assurance of the reality of transcendental space, I strive to bring my faculties of mental vision into accordance with theirs.[23]

Embedded within this gratuitous name-dropping is a circular sort of logic, a finessing of the absence of origin in line with Baudrillard's simulacrum, 'the generation by models of a real without origin or reality' that results in 'a hyperreal'.[24] To understand how the fourth dimension moved from being a figure of speech in analytical geometry to hyperreal hyperspace, we must consider flatland narratives of lower-dimensional spaces, or, what is more appropriately called the dimensional analogy.

The Dimensional Analogy

The dimensional analogy begins as a thought experiment, where the writer asks the reader to imagine a flat or two-dimensional world complete with living, intelligent, two-dimensional beings, in order to then imagine the relationship between our world and a four-dimensional one. The most famous of dimensional analogies is the one expressed by Edwin Abbott in his 1884 novella, *Flatland: A Romance of Many Dimensions*. *Flatland* serves as a useful point of reference – although the first example of a dimensional analogy in print was Gustav Theodor Fechner's semi-comical essay 'Der Raum Hat Vier Dimensionen' in 1846, Abbott's is the most popular (and detailed) treatment of the dimensional analogy within an individual text.

Flatland is divided evenly into two parts. The first part of this text, titled 'This World', develops and represents this two-dimensional world; the second part, titled 'Other Worlds', completes the analogy by exploring the relationship between Flatland and worlds of other dimensions, such as Spaceland, Lineland and Pointland. Thus, the entire text of *Flatland* is dedicated to working out the dimensional analogy. The dimensional analogy is important for two reasons: firstly, because it is a recurring trope in all hyperspace philosophy and popular four-dimensional fiction I have encountered. Indeed, the trope is so familiar to the subject that by 1910, Paul Bold, in his short story 'The Professor's Experiments', had refined it down to a brief explanation from the titular professor:

> Well then, in the first place we exist in a land of three dimensions – length, breadth, height – and we can ordinarily conceive of no extra or fourth dimension. But we can conceive of beings in the *lower* dimensions, and a being in two dimensions would know of length and breadth, and would have no conception of height; planes or plane surfaces would be the limit of his knowledge, and the third dimension would be as unthinkable to him as the fourth

---

23 Sylvester, 'A Plea', 1: 238.
24 Baudrillard, *Simulations*, 2.

dimension is to us. Again, a being in one dimension would only know of length; both breadth and height would be unthinkable. Do you follow?[25]

That the professor is able to relay the dimensional analogy so briefly is a testament to the familiarity of this device by the early years of the twentieth century.[26]

Second, the dimensional analogy is important because for hyperspace philosophers such as Hinton, it is the device through which the spatial fourth dimension is actually *created*. For those who did believe in the material existence of a higher dimension, the dimensional analogy was not only the means by which this idea was communicated; it was an important tool in locating, describing and even experiencing hyperspace. This is why, although it deployed the dimensional analogy and addressed the fourth dimension, *Flatland* is most accurately situated outside of hyperspace philosophy. As was recognized by some of his contemporaries – and more recently by literary critics – Abbott was not as concerned with popularizing the fourth dimension here as he was with satirizing contemporary English culture.[27]

Additionally, some critics have read *Flatland* as a result of the Reverend Abbott's 'opportunistic desire to reconcile science and theology', in utilizing the challenge to scientific materialism offered by the new geometries.[28] As Hinton wrote in 1885, he would have liked to recommend the dimensional analogy of *Flatland* to his readers as an instructive example, but

> turning over its pages again, I find that the author has used his rare talent for a purpose foreign to the intent of our work. For evidently the physical conditions of life on the plane have not been his main object. He has used them as a setting wherein to place his satire and his lessons. But we wish, in the first place, to know the physical facts.[29]

Here Hinton underlined the key difference that he saw between his work and Abbott's: Abbott deployed the analogy of a two-dimensional world to direct the reader's attention to the social conditions of our own, three-dimensional world. Hinton, by contrast, wanted to us to consider 'the physical conditions of life on the plane' as a means to finding a strategy for perceiving, 'perchance a help to the comprehension of a higher life' in 'the mysterious minute actions by which [we are] surrounded' in our three-dimensional world.[30] Hinton used analogy to hypothesize and make observations about

---

25 Bold, 'The Professor's Experiments', 257, original emphasis.
26 See also Manning, ed., *The Fourth Dimension Simply Explained*. The majority of these essays rely – explicitly or not – on Hinton's work. In fact, Einstein used a refined version of the dimensional analogy to explain his own theories the nonscientific reader in 1938. See Einstein and Infeld, *The Evolution of Physics*.
27 For nineteenth-century critics see Tucker, 'Review of "Flatland"'; and Hinton, below; for recent literary criticism that addresses Abbott's use of satire, see Jann's introduction to *Flatland*, vii–xxxiii; and Smith, Berkove and Baker, 'A Grammar of Dissent', 129–50.
28 Valente, 'Transgression', 74. See also Jann, 'Abbott's "Flatland', 473–90.
29 Hinton, 'A Plane World', *Scientific Romances*, 129.
30 Ibid., 156.

nature and the act of perception. Perhaps rather paradoxically, by positioning himself as a scientist and speculative philosopher rather than an author, Hinton used the dimensional analogy as means of experimentation and creation, instead of treating it simply as a tool of description.[31]

This difference is one I will continue to highlight throughout this study: in writing a work of satire, Abbott self-consciously used the dimensional analogy as the foundation for a fiction that is ultimately designed to deflect the reader's attention back outward to the social and cultural struggles of the lived, three-dimensional world. In this sense, *Flatland* relies on 'science' as a foundation for its fiction, performing a function similar to that of much traditional science fiction. As a scientific romancer, Hinton used analogy to create the fiction of the fourth dimension; this is a space that is literally engendered by the manipulation of mathematical symbols. Hinton's fourth dimension is the result of accidental and partial movements of terms across the discourses of algebra, geometry and physics.

We see a similar movement in the 'new' psychology of the second half of the Victorian period: Alexander Bain and Herbert Spencer based their explanations of the functions of the nervous system on an analogy with physical theories of force. These theories were attacked on the grounds that they mistook analogy for fact, and 'refused to accept force as merely a mathematical function devised by physicists to aid understanding of matter in motion'.[32] As Rick Rylance notes, theories supported solely by analogy, such as Bain's, 'have a cogency in principle, but are difficult to sustain in detail'.[33] It is in the attempt to flesh out the details of the dimensional analogy that Hinton's version of the fourth dimension is created. At work here is 'the speculative, argumentatively-extended character of analogy' in which, as Beer observes, 'the arc of desire seeks to transform the conditional into the actual'.[34] In the hands of hyperspace philosophers such as Hinton, the dimensional analogy became a transformational and revelatory device.

However, before we explore Hinton's use of this device, it is necessary to examine the development of the dimensional analogy over the forty years preceding his work. In exploring the development of the dimensional analogy from Gustav Fechner onward, I highlight the hypostasization of the terms of analytic algebra into descriptive geometry, which then led to what Hinton called 'scientific romance'.[35]

Before Hinton: The Fourth Dimension 1846–1880

In his 1846 essay, Fechner wrote: 'One imagines a small, colourful little man who walks around in a camera obscura on the paper; here one has a being that exists in two

---

31 For just a few examples, see Beer, *Darwin's Plots*, 73–96; Bohm and Peat, *Science, Order and Creativity*; Papin, 'This Is Not a Universe'; and Arbib and Hesse, *The Construction of Reality*, 147–71.
32 Smith, 'Physiological Psychology and the Philosophy of Nature', quoted in Rylance, 179.
33 Rylance, *Victorian Psychology*, 180.
34 Beer, *Darwin's Plots*, 79.
35 See also Throesch, 'Nonsense in the Fourth Dimension'.

dimensions'. This two-dimensional being has no comprehension of the extra dimension of space, depth, that extends upward and downward from his photosensitive paper. If the philosophical possibility of a third dimension of space even occurred to this 'little man', he would decide that its material existence was impossible. 'Nevertheless', Fechner remarked, 'there exists this third dimension'.[36] Fechner continued, arguing that this little man is in fact representative of humanity with its three-dimensional prejudices: 'We are only little colourful men and little shadow men in three dimensions instead of two'.[37] As the two-dimensional being in the camera is oblivious to the three-dimensional world that human beings inhabit, Fechner argued, so are humans oblivious to the fourth dimension of space.

Strikingly, as Alexander L. Taylor observed in 1952, Fechner's version of the dimensional analogy anticipates animated film: 'At each moment we have a cross-section of this larger [four-dimensional] reality of which we know nothing, any more than, shall we say, Donald Duck, were he conscious, would know of the world beyond his screen'.[38] Taylor's language here exemplifies how the dimensional analogy functions; replacing the two-dimensional camera manikin with a Disney character, he directly implicates his audience in the analogy, referring to the reader as 'we', the three-dimensional beings, are now part of the fiction. Disturbingly, this analogy also implies the possibility that we, too, are being watched by hyper-beings, something Hinton explicitly addressed in the first series of his *Scientific Romances*.

Fechner was not the only one interested in imagining two-dimensional worlds during the decades before Hinton began writing; in fact, he may have borrowed this idea from fellow German mathematician, Carl Friedrich Gauss.[39] In Gauss's biography, published shortly after his death, Sartorius von Waltershausen recalled that Gauss frequently employed a similar analogy in lectures and conversations. Writing in an 1869 issue of *Nature*, Sylvester noted that Gauss often remarked that 'as we can conceive beings (like infinitely attenuated book-worms in an infinitely thin sheet of paper) which possess only the notion of space of two dimensions, so we may imagine beings capable of realising space of four or a greater number of dimensions'.[40] Henderson identifies Sylvester's article as 'a more direct impetus to the rise of English speculation on the number of

---

36 Fechner, *Vier Paradoxa*, 24, my translation. I do not offer a direct translation here; literally, Fechner asks the reader to imagine 'ein kleines buntes Männchen [a little, coloured man]'. In deviating from the original text, I am trying to clarify Fechner's intentions. He is asking the reader to imagine a 'real', living, two-dimensional character whose total realm of experience consists of the light-sensitive plate within the camera. I speculate he describes the manikin as 'coloured' is in order to render it more lifelike, as opposed to the black-and-white negative image of the contemporary calotype.
37 Ibid., 25.
38 Taylor, *The White Knight*, 90.
39 In a manner similar to the independent, concurrent formulations of the theory of evolution by natural selection developed by Charles Darwin and Alfred Russell Wallace, Gauss, along with Johannes Bólyai and Nicholai Lobachevski, 'discovered' non-Euclidean geometry.
40 Sylvester, 'A Plea', 1: 238.

dimensions of space' than Fechner's; indeed, the dimensional analogy began to appear frequently in British scientific journals during the 1870s.[41]

The dimensional analogy became entangled in debates between empiricists and idealists concerning the psychology of space perception. In 1870, Helmholtz first employed the dimensional analogy in an attempt to clarify the slippage in terminology that was already occurring. Updating this discussion of curved and n-dimensional spaces six years later, he reiterated:

> To prevent misunderstanding I will once more observe that this so-called measure of space-curvature is a quantity obtained by purely analytical calculation and that its introduction involves no suggestion of relations that would have a meaning only for sense-perception.[42]

The language here carefully notes that Helmholtz was speaking only in analytical terms and did not intend to attribute any kind of descriptive value to this example. However, in this same article he challenged Kant's claim that the axioms of Euclidean geometry 'are necessary consequences of an *a priori* transcendental form of intuition', arguing that Kant was incorrect because we are able to represent other coherent and non-Euclidean systems of geometry for various curved spaces, as he had just demonstrated.[43] In this, Helmholtz was clearly attacking the idealists, because – as a result of the new geometries – 'it cannot be allowed that the axioms of our geometry depend on the native form of our perceptive faculty, or are in any way connected with it'.[44]

For Hinton and other proponents of the fourth dimension, the path lay somewhere between Helmholtz's empiricism and Kantian idealism. Rather than discard the Kantian a priori wholesale, Hinton retained the framework: recognition of the fourth dimension was, for him, a means of developing and expanding human consciousness. His hyperspace philosophy was founded on Kant's claim that space is the means by which the mind encounters the real; if true, then conceiving and perceiving higher dimensions would allow the mind to develop higher aesthetic and ethical sensibilities.

However, Hinton challenged Kant's claim

> that complete space [...] has three dimensions, and that space in general cannot have more is built on the proposition that [...] cannot be shown from concepts, but rests immediately on intuition, and indeed, because it is apodictically certain, on pure intuition *a priori*.[45]

Not only did Hinton propose four dimensions, but in his second series of *Scientific Romances*, he raised the possibility of an unlimited number of dimensions.[46]

The complex challenge to the Kantian a priori posited by proponents of the new geometries was aptly described by F. C. S. Schiller in 1896:

41 See also Blacklock, 'Analogy and the Dimensional Menagerie'.
42 Helmholz, 'The Origin and Meaning', 308. See also Helmholtz, 'The Axioms of Geometry'.
43 Helmholz, 'The Origin and Meaning', 314.
44 Ibid., 318.
45 Kant, *Prolegomena*, 40–41.
46 See Hinton, 'Many Dimensions', *Scientific Romances*, 27–44.

> At a cursory glance it might indeed seem as though the new geometry afforded a welcome support to the Kantian position. If Euclidean geometry alone could prove the possibility of synthetic judgements *a priori*, [...] surely now that it is reinforced by two or more sister sciences, a boundless extension of our *a priori* knowledge might reasonably be anticipated. Unfortunately it proves a case of 'too many cooks' [...]. Just as the *de facto* existence of geometry seemed to Kant to prove the possibility of an *a priori* intuition of Space, so the *de facto* existence of metageometry [i.e., non-Euclidean geometries] indicates the derivative nature of an intuition Kant had considered ultimate.[47]

The introduction of geometries based on non-Euclidean spaces causes the Kantian a priori to deconstruct itself, revealing its derivative nature and status as artefact. Schiller wondered whether the outcome of this deconstruction was still too 'inchoate and chaotic for its full significance to be determined'.[48] One way of resolving the chaos would be to replace Kant's three-dimensional apodictic certainty with a four-dimensional analogue. A superficial reading of Hinton, particularly his early writings, might allow one to conclude that he is doing just this.

It would be easy to read Hinton's fourth dimension as simply a tweaking of Kantian idealism, perhaps in response to the threat posed by the new geometries. However, in exploring Hinton's œuvre, we will find something more complex at the heart of his hyperspace philosophy, a – to borrow Rick Rylance's phrase – 'gradual conceptual consolidation of multiple sources'.[49] These sources included not only the new geometries and Hinton's immediate personal acquaintances, but current debates in physics, aesthetics and ethics. These sources and discourses are consolidated and expressed within Hinton's hyperspace philosophy as a particular concern with the gap between external reality and internal experience and the role of the creative will in bridging this gap.

For Hinton the gap between external and internal was intimately intertwined with the question of the relationship between experience and intuition, of – in William James's terminology – 'percepts and concepts'.[50] Like William James, Hinton worked on the assumption that 'percepts and concepts interpenetrate and melt together, impregnate and fertilize each other. Neither, taken alone, knows reality in its completeness'.[51] Concepts, though they may become increasingly abstract, originate in perception, and in order to be truthful (in James's pragmatic sense), they must in turn impact perception in a manner that modifies both. Thus, Hinton treated his fourth dimension as a concept. The problem, of course, was the apparent lack of evidence for the origin of this concept in perception. George Henry Lewes voiced the opinion of many sceptics when he argued that while non-Euclidean geometry (including the fourth dimension) 'may be thoroughly consistent, and ideally true', the manipulation of abstract mathematical symbols, though done logically and consistently, does not support 'the legitimacy of extending any of its conclusions beyond that [abstract] sphere'.[52]

47 Schiller, 'Non-Euclidean Geometry', 178–79.
48 Ibid., 174.
49 Rylance, *Victorian Psychology*, 169.
50 W. James, *Writings, 1902–1910*, particularly 1007–39.
51 Ibid., 1010.
52 Lewes, 'Imaginary Geometry', 197–98.

Lewes also disagreed with the claims of Helmholtz and others 'that because we can conceive a Space in which its axioms would not be truths, the Euclidean Geometry is not [...] necessarily true'.[53] Those mathematicians and scientists – such as Helmholtz – who did utilize the new geometries while maintaining the distinction between analytical and descriptive discourses argued it was possible to conceive of and represent the perceptions of beings confined to a two-dimensional plane because to do so we must simply subtract one of our existing sensations. However, it would be impossible to imagine a fourth dimension in addition to our own, because

> as all our means of sense-perception extend only to space of three dimensions, and a fourth is not merely a modification of what we have but something perfectly new, we find ourselves by reason of our bodily organisation quite unable to represent a fourth dimension.[54]

Hinton did not disagree with the empiricism of Helmholtz as expressed here; rather, he sought to prove that humans can *experience* sensations of higher dimensions. He was not the only thinker to do so.

Physicist and spiritualist Johann Carl Friedrich Zöllner used Helmholtz's work to support his own claims for the existence of four-dimensional space.[55] Zöllner, colleague and friend of Fechner, was also fascinated with the fourth dimension. Influenced by the American medium Henry Slade, Zöllner was convinced that he had found experimental proof of the existence of the fourth dimension of space. Slade, most famous for slate-writing, also performed a series of tricks, one of which involved untying the knots of a cord with fused endings. Slade's ability to untie the knots – seemingly without touching the cord or disturbing the fused endings – convinced Zöllner that he was able to access the fourth dimension of space.[56] Although an English court convicted Slade of fraud in 1876, Zöllner continued to support him and rely upon him for empirical evidence of the existence of the fourth dimension: he published in the British *Quarterly Journal of Science* to this effect in 1878, and his book on the subject, *Transcendental Physics*, was translated into English in 1880.[57]

Although he, too, sought proof of a fourth spatial dimension, Hinton eschewed involvement in Spiritualist and Theosophist debates. He also attempted to give his dimensional analogy more solid scientific grounding as opposed to the anthropomorphic narratives of Fechner, Helmholtz, Abbott and others. While not entirely averse to the fictive potential of speculative analogy (as clearly indicated by the chosen title *Scientific Romances* for much of his work), Hinton wanted to emphasize the scientific nature of his speculative analogies. In his first scientific romance, 'What Is the Fourth Dimension?' (1880), we see a variation on the anthropomorphic dimensional analogy:

---

53 Ibid., 193.
54 Helmholtz, 'The Origin and Meaning', 318–19. Conversely, while making a similar distinction between discussing two-dimensional and four-dimensional worlds, Lewes argued that it is only possible to 'symbolically construct a space of two dimensions'. See 'Imaginary Geometry', 200.
55 See Stromberg, 'Helmholtz and Zoellner'.
56 Staubermann, 'Tying the Knot'.
57 See Zöllner, 'On Space of Four Dimensions' and *Transcendental Physics*.

If there is a straight line before us two inches long, its length is expressed by the number 2. Suppose a square to be described on the line, the number of square inches in this figure is expressed by the number 4, *i.e.*, 2 × 2. This 2 × 2 is generally written 2^2, and named '2 square.'

Now, of course, the arithmetical process of multiplication is in no sense identical with that process by which a square is generated from the motion of a straight line, or a cube from the motion of a square. But it has been observed that the units resulting in each case, though different in kind, are the same in number. [...]

We have now a straight line two inches long. On this a square has been constructed containing four square inches. If on the same line a cube be constructed, the number of cubic inches in the figure so made is 8, *i.e.*, 2 × 2 × 2 or 2^3. Here, corresponding to the numbers 2, 2^2, 2^3, we have a series of figures. Each figure contains more units than the last, and in each the unit is of a different kind. [...] The straight line is said to be of one dimension because it can be measured only in one way. Its length can be taken, but it has no breadth or thickness. The square is said to be of two dimensions because it has both length and breadth. The cube is said to have three dimensions, because it can be measured in three ways.

The question naturally occurs, looking at these numbers 2, 2^2, 2^3, by what figure shall we represent 2^4, or 2 × 2 × 2 × 2[?] We know that in the figure there must be sixteen units, or twice as many units as in the cube.[58]

Hinton's decision to use algebraic symbols to represent lower-dimensional entities rather than flatland creatures is indicative of an attempt to respond to recent scientific debates about the possibility of a spatial fourth dimension. As we will see, he constructed practical mental and physical exercises he hoped would open the human consciousness to the perception of a figure that corresponds to 2^4. He also proposed – though never rigorously developed – ways of detecting the fourth dimension on the micro and macro levels through examining the movements of molecules and stellar bodies. Before he could justify attempts to obtain experimental proof of the fourth dimension, however, he needed to prove that it was possible to imagine it.

Helmholtz had argued that the problem with imagining the fourth dimension was that it was 'not merely a modification of what we have but something perfectly new [and] we find ourselves by reasons of our bodily organisation quite unable to represent a fourth dimension'. Hinton addressed this problem in the first scientific romance, explaining that, when trying to represent 2^4,

> instead of trying to find something already known, to which the idea of a figure corresponding to the fourth power can be affixed, let us simply reason out what the properties of such a figure must be. In this attempt we have to rely, not on a process of touching or vision, such as informs us of the properties of bodies in the space we know, but on a process of thought.[59]

Hinton wanted to use the mind to imagine something entirely new – a possibility denied even to Ruskin's highest imaginative artist. To do this, it was necessary to engage with

---

58 Hinton, 'What Is the Fourth Dimension?', *Scientific Romances*, 9–10.
59 Ibid., 10.

problems of representation, and the relativity of knowledge. This was the first step in Hinton's lifelong project of perceiving the 'new' space of the fourth dimension.

Hinton's Early Influences

Above, I have outlined the debates concerning the new geometries and the growing use of the dimensional analogy to demonstrate the possibility of four-dimensional space in the 1870s; this was the intellectual climate in which Hinton came of age. It is important to consider even more specifically the cultural milieu of Oxford in the 1870s, and growing debates concerning the role of science and aesthetics in education. Two key figures of influence for Hinton during this time were his father, James Hinton, and John Ruskin.

Ruskin knew James Hinton personally, and both men were members of the Metaphysical Society in the early 1870s. James Hinton, who was well known in his own time for his philosophical writings, died unexpectedly when his reputation was at its peak. After his death late in 1875, a contributor to the journal *Mind* lamented:

> His death at a critical period of his life, when he had just attained his long-desired speculative freedom, was a painful shock to his friends; nor could any country least of all our own, well afford to lose so earnest, unencumbered and well-equipped a pioneer in the search for the truth.[60]

Similarly, Ruskin mourned the loss of James Hinton in *Fors Clavigera*, writing of a 'dead friend, [...] who could have taught us much'.[61] Like James Hinton's writings in mysticism and social philosophy, Hinton's lifelong project of perceiving the fourth dimension was a 'search for truth'. Hinton edited his father's posthumous collection of writings, *Chapters on the Art of Thinking*, published in 1879, and there is some overlap between their philosophies. Two themes from James Hinton's philosophical writings emerge as especially important for the younger Hinton: 'lawbreaking' and 'service'. While lawbreaking is perhaps most relevant in understanding Hinton's fascination with the fourth dimension, James Hinton's concept of service was most influential in the hyperspace philosophy that his son developed as a result of his interest in higher space. For now, I focus on lawbreaking, but I return to James Hinton's concept of service later in this chapter.

James Hinton argued that true genius lies in lawbreaking, or in removing artificial limitations that are placed on human beings. In his last writings, he argued: 'Man's worst evil is the false laws he puts on himself; and what he makes them regarding himself. What Christ did for him was to show him how to escape'.[62] Drawing on Romantic individualism and anticipating Nietzsche's revaluation of values, James Hinton's lawbreaking underpinned the free-love philosophy for which he became notorious.[63] Very much a

---

60 Payne, 'James Hinton', 252.
61 Ruskin, *Works*, 29: 67.
62 J. Hinton, *The Law-Breaker*, 24.
63 Edith Ellis goes so far as to cite James Hinton as a precursor to Nietzsche; see her *Three Modern Seers*.

product of late-Victorian culture, James Hinton's lawbreaking was clearly a direct influence on Hinton's desire to move past the 'apodictic certainty' of the three-dimensional limitations of space.

Hinton's challenge to Kant's argument that the three-dimensional nature of space is a necessary absolute truth founded on unmediated human intuition shared similarities with – but was not identical to – Helmholtz and others' attempts to undermine the Kantian transcendental intuition. Taking a strictly empiricist approach, Helmholtz was careful to note that there is no evidence to support a theory of four-dimensional physical space. However, by challenging Kant's apodictic certainties about space perception, empiricists like Helmholtz opened the door for the hyperspace philosophy of Hinton.

At issue again is the migration of ideas from context to context, and the unexpected and unintended meanings which can arise from the fluidity of certain terms. The question, as Jonathan Smith has noted, was one of 'conceivability', and what exactly was meant by that term.[64] Proponents of classical geometry such as Whewell took the idealist position that the axioms of geometry were necessarily true because it was impossible to conceive of their contradiction.[65] When Helmholtz challenged the idealist position by arguing that it was possible to 'represent to ourselves the look of a pseudospherical world in all directions just as we can develop the conception of it', he was aware of the innate difficulties of the vocabulary.[66] 'By the much abused expression "to represent" or "to be able to think how something happens"', Helmholtz explained, 'I understand [...] the power of imagining the whole series of sensible impressions that would be had in such a case'.[67] In spite of this attempt at clarification, however, Helmholtz still confused these terms: to think about 'how something happens' is different from 'imagining' or representing sensible impressions. Lilianne Papin observes that 'in Western languages in particular, the process of thinking is linked to seeing', and this is what makes modern physics so difficult to understand.[68] Hinton's fourth dimension – as a transitional concept developed in the gap between Newtonian and Einsteinian physics – encountered some of the same difficulties. Helmholtz struggled not only with the slippage between thinking and seeing, but also the growing differences between how scientists and philosophers used language.

Idealist philosopher Jan Pieter Nicholas Land took Helmholtz to task for creeping across 'the fatal border' between the discourses of science and philosophy. Land's overall argument against empiricist challenges to the intuitive origins of the axioms of geometry was somewhat tautological; he claimed that

> to demand logical proof for genuine geometrical axioms is a mistake, because every proof must proceed from some ultimate premises, which in this case must concern space. There are no data about space either in logic or arithmetic, but only in our sense-intuition, and precisely the data expressed.[69]

---

64 J. Smith, *Fact and Feeling*, 186.
65 Whewell, *The Philosophy of the Inductive Sciences*, 1: 665.
66 Helmholtz, 'The Origin and Meaning', 318.
67 Ibid., 304.
68 Papin, 'This Is Not a Universe', 1256.
69 Land, 'Kant's Space', 39.

However, what interests me in Land's response to Helmholtz is not his rebuttal of the challenge to Kant's transcendental a priori, but rather his discussion of the slippage between the terms of science and the terms of philosophy:

> We are told of spherical and pseudospherical space, and non-Euclideans exert all their powers to legitimate these as space by making them imaginable. We do not find that they succeed in this, unless the notion of imaginability be stretched far beyond what Kantians and others understand by the word. To be sure, it is easy to imagine a spherical surface as a construction in Euclid's space; but we vainly attempt to get an intuition of a solid standing in the same relation to that surface as our own solids stand to the plane. [...] We may cloak our perplexity by special phrases, saying that only limited strips of the surface can be 'connectedly represented in our space,' while it may yet be 'thought of as infinitely continued in all directions'. The former is just what is commonly understood by being 'imagined,' whereas being 'thought of' does not imply imagination any more than in the case of, say, $\sqrt{-1}$.[70]

The distinction between being 'thought of' on one side and being imagined or represented on the other, is one that Land extended to further his idealist stance. We must learn to distinguish between notions of 'reality' and 'objectivity', Land argued: while these concepts are identical for the scientist (or natural philosopher), they are not so to the idealist philosopher. 'Reality' is the term used to denote that which exists outside of the mind of the perceiver, while the 'object' is the impression that is received by the mind of the reality outside of it. The question that a philosopher must address, Land claimed, is how much the object differs from the real. 'If', he continued, 'it were established beyond all doubt that the "object" and the "real" are one and the same, all examination of such questions and theories would become empty ceremony, and the paradoxes of Idealism absurdities unworthy of our notice'.[71]

Land, like Kant, was not a pure idealist in that he acknowledged that there is *something* outside of mind. The philosopher's interest, he argued, lies in the gap between the perceived object and the real. The scientist, in order to be able to formulate and test hypotheses, must assume that these are one and the same. The empirical method is not applicable to Kant's discussion of space intuition because, Land argued, our experience of space is necessarily filtered through our space intuition, which is a priori. What is interesting here is how Land left open the possibility for the actual existence of a fourth spatial dimension. Since scientists and mathematicians are able to theorize about the properties of four-dimensional space, Land continued,

> there is no reason to deny the same faculty to our imaginary surface-men. [...] Some genius among them might conceive the bold hypothesis of a third dimension, and demonstrate that actual observations are perfectly explained by it. Henceforth there would be a double set of geometrical axioms; one the same as ours, belonging to science, and another resulting from experience in a spherical surface only, belonging to daily life. The latter would express the 'object' of sense-intuition; the former, 'reality,' incapable of being represented in empirical

---

70 Ibid., 41.
71 Ibid., 40.

space, but perfectly capable of being thought of and admitted by the learned as real, albeit different from the space inhabited.[72]

Thus, Land employed his own dimensional analogy, with the implication that four-dimensional space might exist, albeit as form of the real that is not accessible to human intuition. It is therefore unimaginable and unrepresentable, even it if is possible to think and talk about its existence. Here again, is the distinction between Vernunft and Verstand, which Sylvester feared was being blurred by English philosophers.

Such debates about the nature of space were part of a larger cultural divide between idealist and empiricist philosophers; similarly, debates concerning Euclidean and the new geometries were invested with underlying class allegiances. The theory of the fourth dimension became a focal point for these debates during the 1870s. These underlying issues shaped Hinton's hyperspace philosophy, which was undoubtedly informed by his own experiences as an Oxford undergraduate. Hinton began his career at Oxford as an unaffiliated student in 1871, and later joined Balliol College in 1873. Balliol at this time was known for its modern liberalism, as well as its philosophical idealism. Thomas Hill Green, who later became the first professor of philosophy at the University, was a tutor at Balliol while Hinton was a student. Green's lectures on Kant likely influenced Hinton, as there are clear echoes of Green's ideas in Hinton's hyperspace philosophy. During the time that the dimensional analogy was appearing with increasing frequency in British periodicals, Green was lecturing his students on Kant's *Critique of Pure Reason*. In Green's interpretation of Kant:

> The primariness or *a priori* character of the ideas which constitute space and time [...] means that it is the condition, without which no feelings would become outward things, so that all other conditions of 'phænomena' may be supposed absent, but not that. [...] In this lies the explanation of Kant's distinction between the idea of space as an *intuition* and other ideas as *conceptions*.[73]

What is implied here – at least in Hinton's later interpretation of Kant via Green – is that the intuition of space is the condition of all perception. To somehow expand this intuition would therefore be to expand the perceptual capabilities of the mind, and this idea became the foundation of Hinton's hyperspace philosophy, which is itself a project of consciousness expansion founded on a strange blend of constructivism and positivism.

Green was not Hinton's only influence at Oxford. While at Balliol, he was acquainted with Arnold Toynbee, who later became an influential figure for social reformers in the 1880s and 1890s.[74] Hinton was also a member of Ruskin's inner circle of undergraduate followers, working as a captain on the Hinksey road project.[75] In a diary entry for 10 December 1874, Ruskin recorded looking at Turner paintings with Hinton, breakfasting

---

72 Ibid., 40.
73 Green, *Works*, 2: 10–11, original emphasis.
74 See Toynbee, ed. *Reminiscences and Letters*, 177.
75 See Hilton, *John Ruskin*, 252–67 and 287–304.

with his 'Balliol men' and going for walks with various students around this time.[76] In fact, a character in Hinton's 1895 novella, *Stella*, appears to be have been modelled on Ruskin, the elder Victorian sage whom the younger male characters of the novel visit and idolize. Hinton likely attended some of Ruskin's lectures in the early 1870s, and would have certainly been familiar with his earlier work. At this stage in his life, Ruskin was concerned in part with social works, as evinced by his sponsorship of the Hinksey project and his series of pamphlets, *Fors Clavigera*, begun in 1871. In addition to his early interest in drawing, Hinton had expressed an interest in 'studying geometry as a direct act of perception' as early as 1869, and he would have been particularly interested in Ruskin's lectures on the relationship between the arts and sciences in early 1872, in which he claimed 'the sciences of light and form (optics and geometry)' to be in 'true fellowship with art'.[77]

Hinton's later desire to instruct others towards a new way of seeing – as expressed through his hyperspace philosophy – is not dissimilar from Ruskin's work as a critic and teacher. Elizabeth K. Helsinger has noted how 'reading Ruskin can become learning to see with Ruskin', and more recently Francis O'Gorman observed that *Fors Clavigera* 'requires its readers to perceive, to discern truths in a manner of the great artists', as Ruskin originally outlined in the third volume of *Modern Painters*.[78] In his informal tutorials with Ruskin, as well as in more formal lectures and by reading, Hinton would have been introduced to Ruskin's idea of the great artist who is able simultaneously to perceive and keep separate objective and subjective accounts of the outside world.

Although he was to influence a number of second-generation British idealists and Balliol men, Ruskin mocked the English proponents of German idealism in his famous discussion of the pathetic fallacy: 'German dullness, and English affectation', he wrote, have caused the 'objectionable' terms, *objectivity* and *subjectivity*, to be too much in vogue. Ruskin offered his own interpretation of British idealists' use of these terms:

> The qualities of things which thus depend upon our perception of them, and upon human nature as affected by them, shall be called Subjective; and the qualities of things which they always have, irrespective of any other nature, as roundness or squareness, shall be called Objective.[79]

Ruskin proposed simplifying these terms to the 'plain old English' phrases of 'It seems so to me' and 'It *is* so',[80] which elides the empirical gap that fascinated the British idealists: he aligns the objective, or 'It *is* so', with 'the ordinary, proper, and true appearances of things to us', and the subjective to the pathetic fallacy.[81] The conflation of the objective and subjective is obvious here; in simplifying the terms, the appearance of things

---

76 *The Diaries of John Ruskin: 1874–1889*, 830. Ruskin refers simply to 'Hinton' in the diary entry, and Evans and Whitehouse speculate that this is James Hinton. However, given the date and his relationship with Ruskin at the time, I believe this refers to Charles Howard Hinton.
77 See J. Hinton, *Life and Letters*, 251–52. Ruskin, *Works*, 4: 193–94.
78 Helsinger, *Ruskin and the Art of the Beholder*, 3; and O'Gorman, 'Ruskin and Particularity', 130.
79 Ruskin, *Works*, 5: 201–2.
80 Ibid., 203, original emphasis.
81 Ibid., 204.

irrespective of a perceiver collapses into 'true appearances of things *to us*', the subjective perceivers.

The objective and subjective are just the first two classes in Ruskin's hierarchy of perception: the third encompasses both photographic/objective realism and subjective pathos, while managing to distinguish between the two. Ruskin identified this third class as belonging to the 'first order of poets'. Above all three of these modes of perception, however, is a fourth, a sort of passive hyper-perception where

> men who, strong as human creatures can be, are yet submitted to influences stronger than they, and see in a sort untruly, because what they see is inconceivably above them. This last is the usual condition of prophetic inspiration.[82]

It is this fourth way of seeing that Hinton wanted to activate in his readers by training them to 'see' hyperspace. However, in order to not be overcome by such a vision, these hyper-perceivers would need to transcend Ruskin's highest order of poet. This involved a sort of evolution of aesthetic sensibility whereby the perceiver would be able to maintain the clear vision of the first order of poets when presented with something that 'is inconceivably above them'. To instigate this evolution, the inconceivable must become conceivable and the intuition must be prepared through the education of the imagination.

The Ruskinian Imagination

It is instructive here to turn to Ruskin's early writings on the imagination. In the second volume of *Modern Painters*, he wrote that the greatest works of art are not those that mimetically transcribe the real world, but those that 'invariably receive the reflection of the mind' of the artist and 'are modified or coloured by its image'. 'This modification', Ruskin explained, 'is the Work of Imagination'.[83] Ruskin devoted an entire section of this volume to defining and describing the imagination, which he distinguished from conception. Ruskin's definition of conception is important for our understanding of debates between geometers and philosophers regarding the conceivability of the fourth dimension.

Ruskin distinguished between two ways of knowing a material object: one is verbal, whereby certain facts are stored in the brain 'as known, but not conceived', which 'we may recollect without any conception of the object at all'. The other is visual, whereby facts about the object exist in the brain as images, 'which [...] would be difficult to express verbally', or to represent.[84] According to Ruskin, the latter way of knowing an object is conception, but it is still *not* imagination. To say that something is conceivable therefore means one is able to visualize an object but cannot represent it to another person. Only the artist, who possesses the imaginative faculties, is able to conceive of

---

82 Ibid., 209.
83 Ruskin, *Works*, 4: 223.
84 Ibid., 229.

something *and* represent it accurately, to put it out into the world as something to be perceived by another.

The artist in possession of the associative imaginative faculty is able to create a harmonized whole:

> If [...] the combination made is to be harmonious, the artist must induce in each of its component parts (suppose two only, for simplicity's sake), such imperfection as that the other shall put it right. If one of them be perfect by itself, the other will be an excrescence. Both must be faulty when separate, and each corrected by the presence of the other. [...] The two imperfections must be correlatively and simultaneously conceived. This is imagination [...] two ideas which are separately wrong, which together shall be right, and of whose unity, therefore, the idea must be formed at the instant they are seized, as it is only in that unity that either are good, and therefore on the conception of that unity can prompt the preference.[85]

Hinton echoed Ruskin in his second series of *Scientific Romances* when he wrote that 'imagination, acting on perception of the outer world, enables the artist to see exactly how his picture would look if a strip of colour or a new form were introduced'.[86] To perceive such a unity and translate it into a work of art is to be 'an inventor', to enact a 'prophetic action of mind'.[87] Like the scientist, the imaginative artist hypothesizes a potential synthesis of incomplete fragments and proceeds to test that hypothesis. The unseen possibility of this synthesis is the same proposed by non-Euclideans and hyperspace philosophers, whom Land derides as claiming reality for something of which 'only limited strips of the surface can be "connectedly represented in our space"'.[88] An imaginative, 'great' artist is needed to translate the thought into reality, to unify the limited conceivable strips into a perceivable harmonious whole. It is the imagination that allows the artist to reveal the invisible harmony from existing visible fragments. This movement from seen to unseen shares the creative potential of the analogy.

There is a second faculty of the imagination according to Ruskin, which is just as important as the associative: this is a 'penetrating, possession-taking faculty', which clearly presupposes a subjective ego. This faculty allows the imagination to see 'the heart and inner nature' of things.[89] Here we see the desire to obliterate the subjective ego while simultaneously protecting it. The imaginative subject is needed to penetrate the superficial appearances of the material world, while at the same time it must not be led astray by its subjectivity. Jay Fellows observes this paradox in Ruskin when he notes that 'to lose sight of oneself is to become an invisible man. And only the invisible man is worthy of self-portraiture', according to Ruskin.[90] Lindsay Smith, who examines Ruskin's early

---

85 Ibid., 233–34.
86 Hinton, 'On the Education of the Imagination', *Scientific Romances*, 5.
87 Ruskin, *Works*, 4: 233–34.
88 Land, 'Kant's Space', 41.
89 Ibid., 251 and 253.
90 Fellows, *The Failing Distance*, 71. For a discussion of this paradoxical ideal in relation to scientific epistemology in the nineteenth century, see Levine, *Dying*.

writings in relation to contemporary developments in optical technologies and the resultant physiological debates, claims that what Ruskin wanted was

> an observing subject that retains the prerogative of the Romantic wanderer, [...] while incorporating contemporary Victorian developments in visual theory. The result is an inevitably strange hybrid: a desire for an invisible man, a poetic identity who is newly aware visually, but whose intelligence absents itself and whose educated eye avoids self-assertion.[91]

Hinton's fourth dimension similarly functioned as a paradoxical space of self-transcendence and self-possession, as we will see in the next two chapters. While his hyperspace philosophy was no doubt informed directly by Ruskin, both men were participants in what George Levine has identified as 'the epistemological ventures of modernity [which] are thick with paradox – materiality entails the incorporeal, the self gains its power by annihilating itself'.[92]

Such paradoxes preclude simple contrasts between Ruskin, the anti-sensualist on the one hand, and Walter Pater and the Aesthetes on the other. As Nicholas Shrimpton and others have demonstrated, there is no clear-cut opposition possible here. Although in the 1880s Ruskin took care to differentiate between the what he saw as the crass sensualist perception of beauty championed by the Aesthetes ('aesthesis') and his own moral perception of beauty ('theoria'), Shrimpton rightly notes that the difference here was one of degree, not kind.[93] Anticipating the quarrel between H. G. Wells and Henry James over the art of fiction, 'Ruskin's argument with the Aesthetes had the bitterness and intensity often associated with internecine quarrels, and an internecine quarrel is precisely what it was'.[94] Kenneth Daley's work is useful here in his examination of Pater's refiguration of Ruskin's pathetic fallacy, which he claims 'converts what Ruskin judges to be intemperate passion into a heightened sense of sympathy and pity, thereby rescuing what Ruskin condemns in romantic practice'.[95] As I demonstrate in the next chapter, rather than attempt to avoid the pathetic fallacy, in his early *Scientific Romances*, Hinton also attempted to push *through* it toward a heightened, four-dimensional consciousness.

Hinton's hyperspace philosophy can be read in part as a response to Ruskin's influence over Victorian aesthetic debates; his push to develop a higher, four-dimensional consciousness was an attempt to address Ruskin's ideal of the creative power of the associative imagination, which 'seizes and combines at the same instant, not only two, but all the important ideas of a poem or picture, and while it works with any one of them, it is at the same instant working with modifying all in their relations to it, never losing sight of their bearings on each other'.[96] This creative agent, which seems to be made 'after the image of God', is decidedly male, but it must encompass the 'powers' that Ruskin elsewhere attributes to the female: ordering, arrangement, sympathy and

---

91 L. Smith, *Victorian Photography*, 25–26.
92 Levine, *Dying*, 2.
93 Shrimpton, 'Ruskin and the Aesthetes', 138.
94 Ibid., 147.
95 Daley, *The Rescue of Romanticism*, 134.
96 Ruskin, *Works*, 4: 236.

passivity.[97] The manly aesthetic ideal of the penetrative imagination, of the poet who is strong enough to experience passion while maintaining constant self-control, resulted in strains and stresses that manifested themselves in interesting (and tragic) ways in the lives and writings of both Ruskin and Hinton. Hinton's hyperspace philosophy, like Pater's reclamation of the Italian Renaissance, can be read as an attempt to engage with (and re-envision) Ruskin's ideal creative agent.

Hinton's Hyperspace Philosophy

In his hyperspace philosophy, Hinton strove to do something more than simply popularize the theory of the fourth dimension. He proclaimed his interest in geometry at a time when discussions of the new geometries were reaching the British press, and it is clear that from the beginning he was drawn to explore the aesthetic implications of the new trend in geometry. In 1869, James Hinton wrote to his son:

> I am glad you like the idea of studying geometry as an exercise of direct perception. I think it must be specially valuable so; and I am very pleased that you think it practicable and useful. The habit of looking thoroughly and minutely into things, alike with the eyes and with the reason, so as to cultivate the power of *seeing* their qualities and relations, and not merely trying to infer them, must be a most excellent one. It will be most valuable to you.[98]

Apparently, in recent correspondence Hinton had indicated his interest in geometry. In celebrating the importance of looking 'alike with the eyes and with the reason', James Hinton proposed a relationship between percepts and concepts similar to that of William James and, before him, Ruskin. In fact, Hinton's later sympathy with William James may be in part due to the fact that James's philosophy was 'somewhat eccentric in its attempt to combine logical realism with an otherwise empiricist mode of thought'.[99] 'Logical realism' here means 'the platonic [*sic*] doctrine [...] that physical realities are constituted by the various concept-stuffs of which they "partake" '.[100] Hinton's willingness to pursue 'eccentric' combinations of philosophical schools echoes the work of his father as well as Ruskin, and would have appealed to William James.

In the same letter, James Hinton urged his son to consider 'the knowledge of phenomena, that is, of what the senses can perceive, [as] the best basis you can lay' for future studies.[101] Aside from his original transubstantiation of the spatial fourth dimension via analogy, Hinton followed this advice throughout his career. His hyperspace philosophy

---

97 See, for example, Ruskin's lecture 'On Queens' Gardens', in *Works*, 18: 122. Certainly, the difference in aesthetic views of Ruskin and Pater can be read as in some measure informed by sexual orientation and identity. See, for example, Daley, *The Rescue of Romanticism*; and Brake, 'Degrees of Darkness'.
98 Hopkins, ed., *Life and Letters*, 251.
99 W. James, *Writings, 1902–1910*, 1037.
100 Ibid., 1036–37.
101 Hopkins, ed., *Life and Letters*, 251.

was also an eccentric combination that Mark McGurl has aptly described as 'transcendental materialism':

> The particular appeal of the fourth dimension was as a potential means of reintegrating the two sides of [...] the 'Omnipresent Debate' in the nineteenth century between empiricism and transcendentalism, or, more roughly, between the competing cultural authority of science and religion. [...] More broadly, non-Euclidean geometry suggested in its own way the possibility of a 'transcendental materialism' similar in some respects to that being developed by figures such as Walter Pater, whose aestheticism merged the traditions of British empiricism and German idealism.[102]

Indeed, Hinton's hyperspace philosophy should be read as an attempt to provide a new basis for considering matter and spirit; at its foundation it is a celebration of mediation, of the technology of representation. For him, 'space conditions' are the fundamental mediator, as he explained in the opening pages of his first book-length philosophical treatment of the fourth dimension:

> It is generally said that the mind cannot perceive things in themselves, but can only apprehend them subject to space conditions. And in this way the space conditions are as it were considered somewhat in the light of hindrances, whereby we are prevented from seeing what the objects in themselves truly are. [...] There is in so many books in which the subject is treated an air of despondency – as if this space apprehension were a kind of veil which shut us off from nature. But there is no need to adopt this feeling. The first postulate of this book is the full recognition of the fact, that it is by means of space that we apprehend what is. Space is the instrument of the mind.[103]

While Hinton accepted the assumption that space apprehension is 'a kind of veil' between the perceiving mind and reality, he disagreed with the idealist philosopher's interpretation of this 'fact'. It is not a limitation to despair of, he argued. Identifying space as 'the instrument of the mind' opens up new possibilities for the mind; accepting space 'as the instrument of the mind' allows the possibility that – by tuning the instrument – humans can embark on 'a new era of thought'.

Hinton's hyperspace philosophy is aptly described as a kind of materialism because he emphasized the textual nature of his project: for him, there was nothing outside of space perception. Nevertheless, hyperspace philosophy can also be described as transcendental because it proclaims a 'higher' form of mediation out there to be discovered and developed. His was not an absolutist project: as we will see, Hinton was open to the possibility that there were 'many dimensions' beyond four. What was important for Hinton and many modernists who were interested in the fourth dimension was the *process* of realizing the fourth dimension. As Bell and Lland have observed, for Hinton and twentieth-century hyperspace philosophers such as Claude Bragdon, 'the Fourth Dimension means

---

102 McGurl, *The Novel Art*, 62.
103 Hinton, *A New Era*, 2.

not so much an attainable place, but a matter of developed consciousness, a process of exploration'.[104]

Hinton believed that undergoing this process of development would result in an aesthetic and social revolution. Here James Hinton's thoughts on 'service' are particularly important. As a young man, Hinton was encouraged to consider the broader implications of his work and to treat the coming era as portentous. In 1870, James Hinton wrote to his son to congratulate him on his decision to refuse confirmation into the Church of England, which he regarded as a positive step toward making 'all your life transparent' and 'banish[ing] all the false pretences which fill our present life with evil'. James Hinton saw the organized Church as hypocritical and many of its members cynically political. Although in one sense Hinton was being groomed for worldly success (he was at the time a student at Rugby and soon to be an Oxford undergraduate), his real mission, according to his father, was to enable himself 'to take up what we [James Hinton's generation] leave unfinished, and perfect what we do incompletely': to, in a sense, become Ruskin's paradoxical invisible man. This task was of the utmost importance because, according to James Hinton,

> it is a great age of the world for which you are preparing – an age in which the great question of true significance of human life will, at least, begin to decide itself. [...] This is one question men will have to answer, Is it our nature to take the best care of ourselves or to live in giving up? I know how your heart would answer this, and I think the time is coming when all men will give the same.[105]

James Hinton's concept of altruistic 'service', like Ruskin's invisible man, was full of contradictions that Hinton attempted to reconcile within his hyperspace philosophy.

Concerned with the relationship between the material and spiritual, particularly with reference to morality, James Hinton argued that morality needed to be approached in a more 'scientific' manner; while the sciences had embraced inductive reasoning, moral philosophers and theologians were still struggling 'to find a "right" for [...] feelings and [...] actions without having laid the basis of a true response to facts'.[106] Contemporary morality was currently centred on the self, James Hinton argued, citing the frequent opposition of desire for pleasure against the 'goodness' of duty as support for his case. He wrote:

> The thought of goodness in diminished pleasure betrays its origin: it arose from putting self first; which perverts the thought of goodness into that of self-restraint: – into goodness *about* self and for its sake.[107]

According to James Hinton, the basis for morality should be altruism, and by this word he meant 'Myself in and for others'.[108] Here he proposed a kind of self-fulfilment

---

104   Bell and Lland, 'Silence and Solidity', 2: 124.
105   Hopkins, ed., *Life and Letters*, 254.
106   J. Hinton, 'On the Basis of Morals', 782.
107   J. Hinton, *Chapters on the Art of Thinking*, 71.
108   Hopkins, ed., *Life and Letters*, 260.

through surrendering oneself to the needs of others. There was no need to waste intellectual and moral 'strength in efforts to rise above sense'.[109] Rather, James Hinton argued that the perfect moral condition would consist of an alignment of desire with serving others. What drew Havelock Ellis and other progressives to James Hinton's writings was the implication that if personal pleasure was sometimes the outcome of fulfilling others' needs, then pleasure was to be embraced as well.

Hinton saw the study of space as a means of obtaining James Hinton's perfected state of altruistic desire. For the four-dimensional consciousness, the difference between duty and desire dissolves. In the dimensional analogy, we who live in three dimensions are able to see the inner workings of two-dimensional creatures; we can observe underlying unities to which they remain blind. Similarly, Hinton explained that, 'to our ordinary [three-dimensional] space-thought, men are isolated, distinct, in great measure antagonistic'. However, after undergoing the process of realizing a four-dimensional perspective, 'it is easily seen that all men may really be members of one body, their isolation may be but an affair of limited consciousness'.[110] The higher viewpoint is expressed in numerous ways, from the penetrating light of X-rays, to the mystical 'mother-sea of consciousness' of Fechner and William James.[111] It is not surprising that many fin-de-siècle progressives were drawn to Hinton's fourth dimension. Boundaries of class and gender were dissolved under the levelling gaze from the fourth dimension.

The surrender of the self's desires to others' needs is paradoxically self-centred: there is no longer an absolute standard of 'right' and 'wrong'; rather, moral judgement must be made on a case-by-case basis, to be determined by the internal condition of the individual in question. Taking issue with moral philosopher and acquaintance Henry Sidgwick, James Hinton made the case for moral relativism. In his 1874 *Methods of Ethics*, Sidgwick had argued for the 'fundamental assumption' of an absolute standard of right and wrong. However, James Hinton wrote that

> reflection shows us not only that right and wrong are qualities incapable of pertaining to things, inasmuch as the same external deed will be, by universal consent, right or wrong, not only under different circumstances, but according to the feelings prompting it. Thus a father rightly chastises a son for a fault for the son's good; but the same blow given in selfish anger would be a crime. [...]
>
> That which is wrong if done for oneself may become right when the claims of 'good' demand it. And the reason of the paramount importance of this response or non-response of the emotions to facts is obvious; it is a question of truth or falsity, of accord or discord between our consciousness and the world.[112]

Hinton was well aware of the revolutionary implications of his father's reliance on service as a basis for morality. In her testimony at his bigamy trial, Hinton's second wife,

---

109 J. Hinton, 'On the Basis', 785.
110 Hinton, *A New Era*, 97.
111 See W. James, *Writings, 1878–1899*, 1100–27.
112 J. Hinton, 'On the Basis', 782.

Maude Florence, explained that 'he did not marry [her] to hurt anyone else, but simply in order that she might have a certificate for her children'.[113] Of course, these children were (presumably) Hinton's children as well, conceived after his marriage to his first wife, so Hinton's moral justification rings somewhat hollow.

However, the fact that Hinton viewed his bigamous marriage in the light of his father's philosophy of service is supported by his oblique reference to the incident in *A New Era of Thought*, a text which was published in 1888 with a cryptic preface by the editors noting that Hinton had left the manuscript in an unfinished state 'on his leaving England for a distant foreign appointment'.[114] This was in reference to Hinton's flight from Britain after his bigamy conviction. Later, within this book, Hinton highlighted the 'dangerous' nature of his claim that it is necessary to 'cast out the self' in order to access hyperspace.

> The problem as it comes to me, is this: it is clearly demonstrated that self-regard is to be put on one side – and self-regard in every respect – not only should things painful and arduous be done, but things degrading and vile, so that they serve.
>
> I am to sign any list of any number of deeds which the most foul imagination can suggest, as things which I would do did the occasion come when I could benefit another by doing them; and, in fact, there is to be no characteristic in any action which I would shrink from did the occasion come when it presented itself to be done for another's sake. And I believe that the soul is absolutely unstained by the action, provided the regard is for another.[115]

Given the stilted language here, the grammatical awkwardness and obscure referent, it is only comprehensible as an allusion to Hinton's bigamy conviction. In the following chapters we will see how the moral relativity of 'service' played itself out in Hinton's hyperspace philosophy. For the present, it is important to observe Hinton's vision of hyperspace philosophy as a moral endeavour.

Writing of Hinton's hyperspace philosophy, Bruce Clarke identifies it as 'a specific and significant response to the evolutionistic vogue for superhuman types' at the turn of the century.[116] Certainly, in Part Two of this book, I read Hinton's work as part of such a response, alongside Wells, Nietzsche and the James brothers, all of whom were concerned with accessing, developing and liberating a higher aesthetic will. However, it is important to acknowledge the roots of Hinton's project as well: his striving to develop the hyperconscious self is also an attempt to transcend Ruskin's artist of the highest order and to engender the 'lawbreaker', of whom James Hinton wrote. Recuperating Hinton's hyperspace philosophy allows us to better understand its context; his work stands as yet another link between the periods and movements traditionally associated with either the nineteenth or the twentieth centuries, Victorians or moderns.

---

113 'Extraordinary Confession of Bigamy', n.p.
114 A. Boole and Falk, 'Preface', in *A New Era of Thought*, v–vii, v. Alicia Boole, a mathematician in her own right, was also Hinton's sister-in-law.
115 Hinton, *A New Era*, 90.
116 Clarke, *Energy Forms*, 185–86.

Chapter Two

CONSTRUCTING THE FOURTH DIMENSION: THE FIRST SERIES OF THE *SCIENTIFIC ROMANCES*

Roland Barthes noted the transgressive potential of analogy, writing that 'its constitutive movement is that of cutting across'.[1] The relational nature of analogous reasoning is also the essential feature that identifies the *Scientific Romances*: holding it together as a larger text is the concept of the fourth dimension itself. Hyperspace philosophy functions as a site where two different impulses – the 'transcendental materialism' described earlier – are linked. As a 'Text' (in Barthes's sense of the word), Hinton's fourth dimension disrupts the separation between objectivity and subjectivity, placing these two perceptual modes in tension with each other.[2] Hinton's conception of the fourth dimension of space is intrinsically idealist; however, in his 'scientific' arguments he supported his theory along materialist lines. Nowhere is this contradiction more observable than in the final text of the first series, 'Casting Out the Self', where he attempted to find an empirical means of approaching the fourth dimension through introspection. The only way such a conflicted project can survive its own construction is through its functioning as a Text with a significant amount of 'play', again in Barthes's sense of the term:

> 'Playing' must be understood here in all its polysemy – the text itself *plays* (like a door, like a machine with 'play') and the reader plays twice over, playing the Text as one plays a game. [...] The Text [...] asks of the reader a practical collaboration.[3]

In this chapter I examine the ways the overall structure of the *Scientific Romances* requires the reader to work with Hinton in the construction of the fourth dimension. Here I focus on how Hinton explicitly calls upon the reader for 'practical collaboration' in playing his

---

1   Barthes, 'From Work', 193.
2   Barthes defined the 'Text' in opposition to the 'work', writing that 'the difference is this: the work is a fragment of substance, occupying a part of the space of books (in a library for example), the Text is a methodological field' (193). I am aligning transcendence with 'romance' or idealism here, and 'materialism' with Victorian empiricism and scientific discourse. I am informed by Karen Armstrong's discussion of *mythos* and *logos* as opposing worldviews. See Armstrong, *The Battle for God*. I also have in mind Ernst Cassirer's distinction between what he called 'discursive thought' and the myth-making activity of the mind, with its tendency toward pathetic fallacy. See Cassirer, *Language and Myth*.
3   Barthes, 'From Work', 196, original emphasis.

texts as if they are a game, and the (perhaps, for Hinton, unintended) 'play' that such activity necessarily entails.

In the previous chapter I noted that before Hinton could justify attempts to obtain experimental proof of the fourth dimension, he needed to demonstrate that it was possible to imagine it. Because Hinton's methodology is founded in analogy, the fourth dimension can only be represented through a series of relations. My focus in the present chapter is on how Hinton attempted to represent the fourth dimension in his first series of *Scientific Romances* (1884–1886). This series includes individual texts that were published first as pamphlets and later collected into a single volume. These texts differ widely: the first, 'What Is the Fourth Dimension?' is a meditation on the mathematical possibility of a spatial fourth dimension. It is here that we are first given Hinton's dimensional analogy of algebraic symbols and descriptive geometric shapes that I cited in the first chapter. The next text, or 'romance', is a short story titled 'The Persian King; or, the Law of the Valley', and it offers an allegory of Victorian thermodynamics, where the fourth dimension is the implied means of escape from entropic cosmic death. I discuss these texts in some detail in the present chapter.

The next two texts in the first series, 'A Plane World' and 'A Picture of Our Universe', inform my discussion throughout this book, but I do not thoroughly explicate them here. 'A Plane World' is another philosophical meditation that employs a dimensional analogy similar to Abbott's *Flatland*; however, the structure of this text is not a traditional narrative. As noted in the previous chapter, Hinton's aim in 'A Plane World' was different from Abbott's satirical impulse in *Flatland*. Where Abbott used the dimensional analogy to comment on contemporary issues around class, gender and religion, Hinton wished to examine 'the physical conditions of life on the plane'.[4] To this end, 'A Plane World' consists of direct exposition, anecdotes, diagrams and even cut outs of 'two-dimensional' beings for the reader to make use of in representing a plane world. 'A Picture of Our Universe' takes yet another approach to the fourth dimension, discussing it in terms electromagnetism and the ether before turning to an argument in favour of free will, which is relevant to my discussion in Chapter Three.[5]

In addition to focusing on the first two romances of the first series in the present chapter, I conclude by examining the final romance, 'Casting Out the Self'. This text offers instructions for the reader's guidance in performing a series of exercises with 27 wooden cubes. This was the first of Hinton's attempts to 'approach' the fourth dimension through exercises with practical models. In offering a detailed reading of these three texts here, my intention is to demonstrate the relational, ambulatory nature of Hinton's four-dimensional aesthetic. This emphasis on relations is perhaps most obvious in his cube exercises, though it serves a necessary function throughout his work. Just as the process of reading Ruskin can become learning to see with Ruskin, the process of reading Hinton's *Scientific Romances* becomes learning to 'see' four-dimensionally. The only way Hinton could attempt to represent the fourth dimension was through a process of correction

---

4 Hinton, 'A Plane World', *Scientific Romances*, 129.
5 Clarke examines 'A Picture of Our Universe' with reference to D. H. Lawrence's conception of the fourth dimension and modernist treatments of the ether; see *Energy Forms*, 180–92.

and supplementation within the reader's imagination. He described this in 'The Persian King' as 'the Arabic method of description', which is, he explained, 'used for the description of numerical quantities. For instance […] if we are asked the number of days in the year, we answer first 300, which is a false answer, but gives the nearest approximation in hundreds. Then we say sixty […]', and so on. Or, to apply this method more directly to his own prose, Hinton explained:

> Firstly a certain statement is made about the subject to be described, and is impressed upon the reader as if it were true. Then when that has been grasped, another statement is made, generally somewhat contradictory, and the first notion formed has to be corrected. But these two statements taken together are given as truth […] and so on.[6]

Each text in the first series of *Scientific Romances* functions as a partial statement of the whole. The effect of reading all of the texts together is an overt manifestation of aesthetic response that, according to Wolfgang Iser, occurs during the process of reading:

> Whatever we have read sinks into our memory and is foreshortened. It may later be evoked again and set against a different background with the result that the reader is enabled to develop hitherto unforeseeable connections […]. Thus, the reader, in establishing these interrelations […] actually causes the text to reveal its potential multiplicity of connections. These connections are the product of the reader's mind working on the raw material of the text, though they are not the text itself – for this consists of just sentences, statements, information, etc.[7]

Hinton's fourth dimension cannot be fully articulated within any single text. It is only through the creative act of the reader who undertakes the process of 'establishing the interrelations' between the different narrative, discursive and practical 'statements' that the fourth dimension can be represented. Hinton was trying to engage his readers in an act of the Ruskinian imagination.

The initial movement of the reader's imagination through the juxtaposed texts within the first series is a dynamic one, highlighting what Iser called the 'gaps of indeterminacy' which he identified in James Joyce's *Ulysses*:

> Each chapter prepares the 'horizon' for the next, and it is the process of reading that provides the continual overlapping and interweaving of the views presented by each of the chapters. The reader is stimulated into filling the 'empty spaces' between the chapters in order to group them into a coherent whole.[8]

The reader of the *Scientific Romances* must overlap and connect the differing perspectives on the fourth dimension presented in each individual romance in order to imagine the fourth dimension. To read and understand one text to is to perhaps *conceive* of the fourth

---

6 Hinton, 'The Persian King', *Scientific Romances*, 54–55.
7 Iser, *The Implied Reader*, 278.
8 Iser, 'Indeterminacy', 39.

dimension, but Hinton wanted to take his reader further than this. In filling the gaps between texts and establishing the relations, the reader generates 'the virtual dimension of the text, which endows it with its reality'.[9] An act of strenuous imagination on the part of the reader would allow them to *perceive* the fourth dimension.

To better understand how and why Hinton sought to engender this aesthetic response through the formal aspects of his work, we need to further examine the content of his early writings on the fourth dimension. With this in mind, I turn to three significant romances from the first series, 'What Is the Fourth Dimension?', 'The Persian King' and 'Casting Out the Self'.

'What Is the Fourth Dimension?'

This is Hinton's first known publication on the fourth dimension, published in *Dublin University Magazine* in 1880; it was later reprinted as the first pamphlet in the first series of the *Scientific Romances* in 1884, and as part of the complete series in 1886.[10] Formally speaking, 'What Is the Fourth Dimension?' is the most traditional of all the texts in the first series. It is a philosophical meditation on the possibility of conceiving of a fourth spatial dimension. In its simple, straightforward style we can observe Hinton's initial assumption about the transparency of language, that his fourth dimension could be expressed in the traditional essay form.

This text is well-positioned as the first in the series; chronologically it is an early writing in Hinton's career, and his early attempt to explain the fourth dimension within the formal limits of the essay mirrors the needs of his readers to begin with a simple explanation of this new idea. The reader thus begins the series with a question ('What is the fourth dimension?') and each text in the series serves as an experimental attempt to find an answer. There is a provisionality to 'What Is the Fourth Dimension?' that sets the tone for the entire first series.

Writing in another context, Ian F. A. Bell notes the similarities between the analogy as a methodology of science and technique of modernist poetry: to acknowledge analogy as an instrument of exploration is to rely 'upon a yoking together of conceptual dissimilarity and relational agreement, laying the ground for revised notions of difference that could be both objectively and speculatively exploratory'.[11] The very texture of the first series of *Scientific Romances* is relational; by 'playing' the text, the reader brings together dissimilar but corrective representations of the fourth dimension. The interrogative mode established in the title of 'What Is the Fourth Dimension?' sets the tone for

---

9 Iser, *The Implied Reader*, 279.
10 I write that this is the first 'known' Hinton text on the fourth dimension because during my research I have uncovered another, unsigned essay from 1878 titled 'The Mystery of the Fourth Dimension', also in the *Dublin University Magazine*. The title echoes James Hinton's most popular work, *The Mystery of Pain*; additionally, the theory of the fourth dimension explicated here and the style of writing lead me to conclude that it is highly likely that Hinton authored this text as well. However, because of the ambiguity concerning the authorship, and its lack of inclusion in the *Scientific Romances*, I will not address it here.
11 I. Bell, 'The Real', 121.

the reader, who is to undergo an aesthetic response that mimics the process of scientific discovery through hypothesis and experimentation.

The reader must traverse the individual texts of the first series, establishing the relations that will allow him or her to perform a creative act of imagination: conceiving of the fourth dimension itself. For Hinton, the ethical importance for this aesthetic response was that it resulted in an analogous act of self-creation. As Bell explains, 'to be creative, then, is to explore, to question [...]. To question is to raise the possibility of moving from one state to another. It is always a liminal activity where the self is poised for change at the edge or boundary of things'.[12] The four-dimensional self that will (hopefully) emerge after reading the first series is self-directed and self-creating: an artist of a higher order. Hinton first pointed to this possibility at the conclusion of 'What Is the Fourth Dimension?', where he stated that, aside from the intrinsic interest of the subject of the fourth dimension,

> speculations of this kind [...] have considerable value; for they enable us to express in intelligible terms things of which we can form no image. They can supply us, as it were, with scaffolding, which the mind can make use of in building up its conceptions. And the additional gain to our power of representation is very great.[13]

Here Hinton maintained the difference between the ability to conceive of something and the ability to imagine it. However, being able to conceive or describe a thing is the first step in imagining it; this results in a significant 'additional gain to our power of representation'. Here we see Hinton's tentative movement toward a proposal of a four-dimensional aesthetic, the groundwork for which he was laying in the first series. The individual texts work as the scaffolding; it is the reader's task to develop their own power of representation through the process of making the connections between these texts.

Establishing this scaffolding is a necessary part of the process of formulating the rules for seeing four-dimensionally. According to Raymond Williams, 'the normal process of perception [...] can only be seen as complete when we have interpreted the incoming sensory information either by a known configuration or rule, or by some new configuration which we can try to learn as a new rule'.[14] Hinton was trying to establish a new rule for seeing through description by asking his readers to 'think of' how four-dimensional objects would appear. He deployed the dimensional analogy to substantiate an answer to the question: 'By what figure shall we represent 2^4[?]'. In his answer, Hinton described the pattern of progression from a line to a square to a cube, arguing that in this way we can extrapolate the geometric properties of a four-dimensional object. In 'working in accordance with the analogy' it is possible 'just as by handling or looking at it, [...] to describe a figure in space, and so by going through a process of calculation it is within our power to describe all the properties of a figure in four dimensions'.[15]

---

12 Ibid., 121.
13 Hinton, 'What Is the Fourth Dimension?', *Scientific Romances*, 31.
14 R. Williams, *The Long Revolution*, 39.
15 Hinton, 'What Is the Fourth Dimension?', *Scientific Romances*, 15.

What Gillian Beer has aptly called 'problems of description in the language of discovery' are particularly foregrounded within 'What Is the Fourth Dimension?'; indeed, they underpin all of Hinton's writing.[16] In describing a new idea or something that is unobservable, the scientist must use terms that are already known. Ruskin described this as the 'limit to the power of all human imagination': 'no human mind has ever conceived' something entirely new.[17] Both scientist and artist can only describe something analogically or metaphorically: as N. Katherine Hayles observes, 'a completely unique object, if such a thing were imaginable, could not be described. Lacking metaphoric connections, it would remain inexpressible'.[18] For Hinton, the problem was to bring the fourth dimension into the realm of the expressible. He found two ways of doing this within the *Scientific Romances*: by pushing 'through' the limitation of the metaphorical nature of language to expose its polysemous creative potential, and by marrying form with content.

This marriage of form and content can be observed in the texture of continual supplementation and correction between the individual texts of the *Scientific Romances*. In the later texts of the first series, Hinton offered various ways of answering the question of the fourth dimension that is raised in the first text. In doing so, Hinton directly challenged what Beer calls 'the naïve positivistic equivalence between object and event, or utterance, [that] presupposes a single necessary theoretical outcome'.[19] The range of the appeal of his hyperspace philosophy amongst Hinton's contemporaries was in part due to its openness – its scope for 'play' – particularly in the first series. Hinton's decision to work out his theories within the hybrid genre of the 'scientific romance' indicates his awareness that, as Beer observes,

> language is a heuristic tool, but it may best function at the frontiers of scientific knowledge by adopting a mode which sounds strangely belletristic. Severe one-to-one equivalence may prove to be paradoxically less exact as a working tool than the larger term during the period of theory formation.[20]

Hinton needed to keep his language multi-vocal in the first series because he was trying to counteract the narrowing of vision within current scientific theory. 'What Is the Fourth Dimension?' begins with the proclamation that 'at the present time our actions are largely influenced by our theories. We have abandoned the simple and instinctive mode of life of the earlier civilisations for one regulated by the assumptions of our knowledge'.[21] Increasing specialization of separate scientific disciplines meant it was crucial to not lose sight of 'the constitution of the knowing faculty, and the conditions of knowledge'.[22] Here, again, we see the influence of James Hinton's lawbreaker: in 'What Is the Fourth Dimension?' we are asked to question 'whatever seems arbitrary

---

16 Beer, *Open Fields*, 149–72.
17 Ruskin, *Works*, 4: 236.
18 Hayles, *Chaos Bound*, 31.
19 Beer, *Open Fields*, 157.
20 Ibid., 157–58.
21 Hinton, 'What Is the Fourth Dimension?', *Scientific Romances*, 3.
22 Ibid., 5.

and irrationally limited in the domain of knowledge'. The particular limitation that 'we must suppose away' is – for Hinton – our understanding of space as limited to three dimensions only.[23]

Hinton's interest in the fourth dimension was underpinned by what Alice Jenkins identifies as a Romantic understanding of abstract space: 'not space imagined as a thing, but the *condition* for imagining things'.[24] Jenkins's designation of 'abstract' as opposed to the material space of geographic, social and political practices is most fitting for Hinton's project, particularly in the first series. Critical explorations of the material spaces listed above are, as Jenkins observes, more appropriately described as explorations of place. The fourth dimension of Hinton's hyperspace philosophy is no place, but rather the space of undiscovered possibility. Thus Hinton wanted to keep that space open for as long as possible because 'whatever pursuit we are engaged in, we are acting consciously or unconsciously upon some theory, some view of things'.[25] Hinton struggled to overcome the problem that the three-dimensional 'view of things' causes by limiting our ability to imagine a four-dimensional object.

Were the four-dimensional analogue for a cube to pass through our three-dimensional space, Hinton explained, 'it would seem to us like a cube'. We must have recourse to the dimensional analogy here: 'to justify this conclusion we have but think of how a cube would appear to a two-dimensional being. To come within the scope of his faculties at all, it must come into contact with the plane in which he moves'.[26] Thus, the cube would appear to the flatlander as a square. 'So, to form an idea of a four-dimensional figure, a series of solid shapes [...] has to be mentally grasped and fused into a unitary conception'.[27] To see four-dimensionally is, therefore, to see as does Ruskin's imaginative artist, who is able to fuse together multiple ideas that are 'separately wrong' but, in unification, 'will be beautiful'.[28]

For Hinton, the three-dimensionally limited imagination is only able to perceive four-dimensional objects as a series of three-dimensional 'slices', just as the being limited to two dimensions would only be able to 'see' a two-dimensional slice of a cube. In combining these slices,

> we should have to imagine some stupendous whole, wherein all that has ever come into being or will come co-exists, which passing slowly on, leaves in this flickering consciousness of ours, limited to a narrow space and a single moment, a tumultuous record of changes and vicissitudes that are but to us. Change and movement seem as if they were all that existed. But the appearance of them would be due merely to the momentary passing through our consciousness of ever existing realities.[29]

---

23 Ibid., 5.
24 Jenkins, *Space*, 152, original emphasis.
25 Hinton, 'What Is the Fourth Dimension?', *Scientific Romances*, 4.
26 Ibid., 16.
27 Ibid., 17.
28 Ruskin, *Works*, 4: 233.
29 Hinton, 'What Is the Fourth Dimension?', *Scientific Romances*, 24.

Taken in the context of fin-de-siècle anxieties surrounding evolution and entropy, progress and degeneration, this statement is both comforting and stifling. While nothing is irrevocably lost, the future is already determined. For Hinton, the experience of time is an illusion: it is simply the way that our three-dimensionally limited consciousness encounters the fourth dimension. Like Ruskin – who argued that the highest form of imaginative genius is one that 'seizes, at the same instant', all the incomplete components that can be successfully fused together to form a beautiful whole – for Hinton, the four-dimensional consciousness perceives all that ever was or will be, simultaneously.[30] Temporal and spatial omniscience, like the ability to create something from nothing, is the province of the divine. 'Imagination', Ruskin argued, 'is neither to be taught, nor by any efforts to be attained'.[31] This is where Hinton departed from his early mentor: his hyperspace philosophy was founded on the belief that by expanding the condition by which we are able perceive and conceive (i.e. space), we are able to 'educate' and expand the imagination.

Hinton's explanation of the lower-dimensional being's experience with a higher-dimensional object as an encounter with a series of 'slices' sets the stage for the reader of the *Scientific Romances*. Each text within this series functions as a single 'slice' that, when combined and fused together by an act of imagination, will give the reader a more complete picture of Hinton's fourth dimension. Here is where form expresses the content of Hinton's writing: the act of reading the series thus becomes a creative act of representing the fourth dimension. According to Hinton,

> when [this] faculty is acquired – or rather when it is brought into consciousness, for it exists in every one in imperfect form – a new horizon opens. The mind acquires a development of power, and in this use of ampler space as a mode of thought, a path is opened by using that very truth which, when first stated by Kant, seemed to close the mind within such fast limits.[32]

Bruce Clarke rightly observes that 'ultimately, then, for Hinton the fourth dimension of space […] was not a discovery waiting to be confirmed so much as a creative ideal – a cultural prophecy and moral goal to which the mind ought to be conformed'.[33] Thus the fourth dimension is not a particular place to be discovered; it is the aesthetic and moral potential of each human consciousness.

In the first series, there is certainly a looseness to Hinton's impressionistic system of supplementation, which indicates some amount of resistance to insistence on point-by-point equivalence between signifier and signified. This fluidity would make sense, considering the period in which Hinton was writing; however, to stop here would be to simplify the matter. As Andrea Henderson has demonstrated, even though

> literary critics and art historians have noted […] the influence of late Victorian mathematical developments on modernist conceptions of space […], the development of non-Euclidean

---

30 Ruskin, *Works*, 4: 234.
31 Ibid., 233.
32 Hinton, *A New Era*, 6–7.
33 Clarke, *Energy Forms*, 185.

geometry had an earlier and more fundamental influence on developments in aesthetics because [...] to relinquish faith in the representational powers of classical geometry was to recognize that symbolic systems generally, while they might be characterized by internal coherence, might not be amenable to a 'real interpretation'.[34]

Hinton was in a strange position because, while the rise of non-Euclidean geometries created space in the public imagination for his hyperspace philosophy, his concept of the fourth dimension was the result of pushing *through* analogy for a 'real interpretation' of the symbol x^4. Thus, while in his first series he played upon the semantic looseness of 'the larger term' of the 'fourth dimension', the very concept of the fourth dimension itself is founded on an insistence on seemingly transparent representation taken to the extreme of hypostasization.

Hinton came to the conclusion that we are four-dimensional beings by getting further entangled in the logic of his own dimensional analogy. In using the dimensional analogy to conceive of two-dimensional plane beings, we are actually just imagining very thin three-dimensional creatures: 'If we consider beings on a plane as not mere idealities, we must suppose them to be of some thickness'.[35] Lines and planes are simply abstractions, and so the imagined plane beings would need to have some kind of – probably miniscule – thickness in the third dimension. By finessing the limitations of the dimensional analogy, Hinton made an important cognitive leap here: if there is a fourth spatial dimension, then we, as three-dimensional beings are either merely abstractions in the mind of a hyperbeing, or we have a four-dimensional existence ourselves. Hinton found the first possibility unsatisfactory; he compared it with Berkeley's religious idealism and concluded that 'it is somewhat curious to notice that we can thus conceive of an existence relative to which that which we enjoy must exist as a mere abstraction'.[36] Again, for Hinton, we are saved by our imaginations: if we are able to conceive of a higher dimension, then it is likely that we are of that higher dimension ourselves. However, the implication that we are unconsciously four-dimensional was not enough to completely abolish the underlying fear of being reduced to powerless abstraction of the mind of a hyperbeing. This is a concern that haunts Hinton's hyperspace philosophy, particularly in the first series.

Victorian Thermodynamics and the Fourth Dimension

Clarke writes that, in 'The Persian King', Hinton 'gave the second law [of thermodynamics] the slip by yielding to it so perfectly that it turned into the enabling condition of all motion and thus all energy'.[37] Just as we have seen in 'What Is the Fourth Dimension?', in this second romance we can observe the 'dialectics of subversion and

---

34 A. Henderson, 'Math for Math's Sake', 457.
35 Hinton, 'What Is the Fourth Dimension?', *Scientific Romances*, 30.
36 Ibid., 31.
37 Clarke, *Energy Forms*, 119.

support' underpinning Hinton's hyperspace philosophy.[38] Before further analysing Hinton's methodology of 'pushing through', it is necessary to understand the movement of the fourth dimension from the context of the new geometries to Victorian thermodynamics.

In the second half of the nineteenth century a growing acceptance of the theory of evolution – finally given scientific authority by the publication of Darwin's research – also necessitated the acceptance of irreversible change and loss. Irreversible change, when applied to thermodynamics, results in the concept of entropy. The second law of thermodynamics states that a certain amount of heat is lost in every exchange of energy; it is not destroyed, but it becomes inaccessible for further exchange. It is this difference that allows for the transfer of heat at all: heat radiates from bodies of higher temperature to those of lower temperature until equilibrium is reached, and this process is irreversible. Eventually, according to late-Victorian physicists, the entire universe would suffer a cosmic 'heat death', a final equilibration.

In 'The Persian King', Hinton never explicitly mentioned the fourth dimension; rather, he created an allegory about entropy. These two concepts were first linked a decade earlier in an 1875 text by physicists Balfour Stewart and Peter Guthrie Tait, in *The Unseen Universe; or, Physical Speculations on a Future State*. This text was widely popular: it was in its fourth edition within a year of initial publication. Stewart and Tait attempted to bridge the expanding rift between the discourses of science and religion; *The Unseen Universe* was written as a challenge to the materialist school of science, and specifically in response to John Tyndall's 1874 Belfast Address to the British Association, in which he argued that the relation of religion to scientific culture was 'grotesque', and that religious irrationality, 'if permitted to intrude on the region of *knowledge*, over which it holds no command' would be a 'mischievous' and 'destructive' force.[39]

Stewart and Tait were particularly concerned with challenging the materialist denial of life after death:

> Take away all hope of a future state, – appear to demonstrate, if not with absolute certainty, yet with an approach to it, that such a condition of things is antagonistic to well-understood scientific principles, and we feel certain that the effect upon humanity would be simply disastrous.
>
> [...] We attempt to show that we are absolutely driven by scientific principles to acknowledge the existence of an Unseen Universe, and by scientific analogy to conclude that it is full of life and intelligence – that it is in fact a spiritual universe and not a dead one.[40]

The thought that there is nothing beyond the material, visible universe was particularly disturbing in the second half of the nineteenth century, after William Thomson's publication of 'On the Universal Tendency in Nature to the Dissipation of Mechanical Energy' (1852), in which he identified entropic implications of the second law of

---

38 J-J. Lecercle, *Philosophy of Nonsense*, 134. See also Throesch, 'Nonsense in the Fourth Dimension'.
39 Tyndall, *Fragments*, 382, original emphasis.
40 Stewart and Tait, *The Unseen Universe*, 3 and 5.

thermodynamics.[41] In one attempt to demonstrate how this seemingly inevitable fate might be avoided, James Clerk Maxwell proposed a thought experiment in which a 'neat-fingered being' is able to reverse the slide into entropic disorder.[42] Maxwell first described this being – later named 'Maxwell's Demon' by William Thomson – in his correspondence with Tait. In one letter, Maxwell asked Tait to imagine a container of gas molecules, divided into two chambers; one has a higher level of thermal energy than the other. The demon is located within the closed container, at the dividing partition, and is able to sort through the molecules by opening and closing a door within the partitioning wall. By sorting the molecules so that those with lower thermal energy pass into the warmer chamber, the demon is able to violate the second law of thermodynamics. As Clarke observes, 'Thomson's name "demon" was an inspired rhetorical choice'.[43] The ambiguous figure of the demon – an unholy divine agent both intermediate and intermediary – is, in the case of Maxwell's 'neat-fingered being', a useful fiction engendered by the desire to re-imagine entropic process in terms other than those of theological and economic dissolution.

In *The Unseen Universe*, Stewart and Tait rejected Maxwell's thought-experiment and its implications. Instead of attempting to save the material universe, Stewart and Tait dwelled on its inevitable and total destruction:

> It thus appears that at each transformation of heat-energy into work a large portion is degraded, while only a small portion is transformed into work. So that while it is very easy to change all of our mechanical or useful energy into heat, it is only possible to transform a portion of this heat-energy back again into work. After each change too the heat becomes more and more dissipated or degraded, that is, less and less available […].
>
> But while the sun thus supplies us with energy he is himself getting colder, and must ultimately, by radiation into space, part with the life-sustaining power which he at present possesses. Besides the inevitable cooling of the sun we must also suppose that owing to something analogous to ethereal friction the earth and the other planets of our system will be drawn spirally nearer and nearer to the sun, and will at length be engulfed in his mass. […] At length, however, this process will have come to an end, and he will be extinguished until, after long but not immeasurable ages, by means of the same ethereal friction his black mass is brought into contact with that of one or more of his nearer neighbours.
>
> Not much further need we dilate on this.[44]

However, Stewart and Tait cannot seem to leave the subject alone: they continued into the next section to ponder the

> mighty catastrophes due to the crashing together of defunct suns – the smashing of the greater part of each into nebulous dust surrounding the remainder, which will form an

---

41 See also Clarke, 'Dark Star Crashes', and Beer, *Open Fields*.
42 Maxwell, *The Scientific Letters*, 2: 332.
43 Clarke, 'Allegories', 69.
44 Stewart and Tait, *The Unseen Universe*, 126–27.

intensely heated nucleus – then, possibly, the formation of a new and larger set of planets with a proportionately larger and hotter sun, a solar system on a far grander scale than the present. And so on, growing in grandeur but diminishing in number till the exhaustion of energy is complete, and after that eternal rest.[45]

Stewart and Tait needed to remind their readers of the threat of impending cosmic death as the result of universal degeneration via entropic dissipation. While, as Hayles explains, much of the rhetoric of physical science writing in the second half of the century contained implicit 'connotations that link these scientific predictions with the complex connections among repressive morality, capital formation, and industrialization in Victorian society', Stewart and Tait made the perceived moral and theological implications of entropy explicit.[46] We live in a fallen *material* universe, Stewart and Tait argued, and nothing can reverse this. Here 'the physical concepts of dissipation and equilibration [are] infused with the moral contents of sin and death'. Stewart and Tait acknowledged Maxwell's demon in their text, but they managed to 'evade the threat it posed to their particular moralization of thermodynamics'.[47]

The only hope for redemption, Stewart and Tait argued, is in another, immaterial or quasi-material, 'unseen universe', the existence of which is supported by 'scientific analogy' and by what Stewart and Tait referred to as the 'Principle of Continuity'. Creating an analogy that is loosely based on Thomson's ring-vortex theory, Stewart and Tait hypothesized the existence of an unseen universe:[48]

> Let us begin by supposing an intelligent agent in the present universe [...] to be developing vortex rings – smoke rings let us imagine [...] just as the smoke-ring was developed out of ordinary molecules, so let us imagine ordinary molecules to be developed as vortex rings out of something much finer and more subtle than themselves, which we have agreed to call the invisible universe.[49]

By the fourth edition of their text in 1876, this unseen universe had become located in the fourth dimension. Unlike the anthropomorphic dimensional anthology that suggests we look for beings similar to ourselves in another, more limited plane of existence, Stewart and Tait proposed examining the movement of particles at the submolecular level. There was nothing, they contended, in scientific theory to disprove the existence of a 'finer and more subtle' invisible universe. Thus, on the assumption that it does in fact exist:

---

45 Ibid., 127–28.
46 Hayles, *Chaos Bound*, 39.
47 Clarke, 'Allegories', 76 and 82.
48 After observing the movement of smoke rings in Tait's laboratory experiments, Thomson formulated a theory of 'vortex atoms', emphasizing their indestructibility. Thomson's work in this area has been cited as a precursor to string theory, the Theory of Everything and knot theory. See Silver, 'Knot Theory's Odd Origins'.
49 Stewart and Tait, *Unseen Universe*, 217–18.

so we may suppose our (essentially three-dimensional) matter to be the mere skin or boundary of an Unseen whose matter has *four* dimensions [...] but may itself consist of four-dimensional boundaries of the five-dimensional matter of a higher Unseen, and so on.[50]

This is the 'redemption' of the visible universe in which we currently reside: 'We shall be led to a universe possessing infinite energy, and of which the developing agency possesses infinite energy'.[51] The fact that the word 'energy' appears so often in this text is itself indicative of late nineteenth-century anxieties concerning the heat death of the universe. Here the fourth dimension is a transcendent, quasi-material realm that functions as a safety net for dissipated energy lost to the visible universe.

'The Persian King; or, the Law of the Valley'

Stewart and Tait's unseen universe allowed them to argue that life after death was not scientifically impossible, while maintaining that the material, visible universe was coming to an end, with all the theological and moral implications that entailed. While Hinton's narrative of the Persian king was clearly influenced by Stewart and Tait's text, he was not invested in shoring up Christian theology here. 'The Persian King' is about entropy and, using the fourth dimension, Hinton found another way around the second law of thermodynamics by 'reversing its moral polarity'.[52] In this story, the dissipation of energy is read as the necessary condition for all life and creativity.

'The Persian King' is divided into two parts. The first part is a narrative about a Persian king who becomes trapped in a remote valley of his kingdom and, it is implied, dies. In this otherworldly, isolated valley, the king is approached by a mysterious old man named Demiourgos, who describes himself as 'the maker of men'.[53] Demiourgos provides the king with a pair of child-like beings to supervise. The children remain inert until Demiourgos instructs the king on how to control them by manipulating their physical sensations:

> [Demiourgos] explained to the king how it could be possible to stimulate the children to activity, for he showed him how he could divest anything that was done of part of its pain and render it more pleasurable than painful. 'In this way thou canst lead the beings I have given thee to do anything,' said the old man, 'but the condition is that thou must take the painful part that thou sparest them thyself'.[54]

In Hinton's narrative, the Demiurge, the Platonic creator of the world, is merged with the limited Demiurge of Gnostic philosophy, which is described as a craftsman who assists in creating the world by fashioning the raw materials provided by the Supreme

---

50 Ibid., 220, original emphasis.
51 Ibid., 220.
52 Clarke, *Energy Forms*, 111.
53 Hinton, 'The Persian King', *Scientific Romances*, 35.
54 Ibid., 39.

Being.⁵⁵ The beings Hinton's Demiourgos produces are limited because they are subject to 'a law […] which binds them in sleepfulness and powerlessness'.⁵⁶ In Hinton's 'fallen' material universe, it is the king who must stimulate the beings provided by Demiourgos by bearing a portion of their pain.

In the second part of this romance Hinton explains the narrative in relation to thermodynamics, electrodynamic field theory and the concept of the luminiferous ether. However, even in the first part of the story, Hinton frequently interrupts the narrative to explain the physics concepts he is using. It is clear early on in the narrative that Hinton is using the pleasure and pain dynamics of the valley as an allegory for thermodynamics: 'The smallest particle there is in the valley lies […] without motion. Each particle has the power of feeling pain and of feeling pleasure, but by the law of the valley these are equal. Hence of itself no particle moves', Demiourgos explains.⁵⁷ Every activity undertaken by the valley-dwellers entails undertaking an equal amount of pleasure and pain; so long as these sensations remain equal, the valley beings will remain in a state of inactivity. By absorbing a fraction of the painful part of the activity, the king disturbs the deadening equilibrium, stimulating the valley-dwellers into action.

The king devises a system for directing these valley-dwellers in increasingly complex movements and activities by combining simpler, repetitive actions:

> As the type of fundamental activity, he chose an action and made the being go through it again and again. Thus the being would go through the act A, then act B. When the action AB was complete it would go through an act of the kind A again, then through an act of the kind B. Thus the creature would be engaged in a routine of this kind, AB, AB, AB, and so on.⁵⁸

Here Hinton relied on an associationist psychological understanding of the nervous system. According to Victorian associationist psychologist Alexander Bain, human brains are educated by performing a set of actions in quick succession so that they eventually become automatically and physiologically linked, such as the kind of coordinated actions one performs when walking or playing a musical instrument. In his influential text, *The Senses and the Intellect* (1855), Bain wrote:

> A stream of conscious energy, no matter how stimulated, causes a muscular contraction, a second stream plays upon another muscle; and the fact that these currents flow together

---

55 See Herbermann et al., ed., *The Catholic Encyclopedia*, 4: 707–8. Clarke makes the connection between Hinton's use of Demiourgos and Plato's *Timæus*, in *Energy Forms* (178). According to the *OED*, 'Demiourgous', 'Demiurgus' and 'Demiurge' are all variables of the same Greek word, defined as 'a name for the Maker or Creator of the world, in the Platonic Philosophy; in certain later systems, as the Gnostic, conceived as a being subordinated to the Supreme Being and sometimes the author of evil'. See *OED Online* [accessed 21 May 2014]. Literally translated, Demiourgos means 'public worker', and 'was originally used to designate any craftsman plying his craft or trade for the use of the public' (Herbermann et al., 707).
56 Hinton, 'The Persian King', *Scientific Romances*, 38.
57 Ibid., 42.
58 Ibid., 56.

through the brain is sufficient to make a partial fusion of the two, which in time becomes a total fusion, so that one cannot be commenced without the other commencing also.[59]

According to this physio-psychological approach, the commands for two separate bodily actions can become 'fused' within the brain so that performance of one action will always trip the other, like a switch. Associationist psychology is underpinned by a mechanistic, utilitarian understanding of human physiology and psychology. Drawing on Jeremy Bentham's theory that sentient beings are primarily driven by the opposing sensations of pleasure and pain, Bain argued that 'pain is what we avoid, repel, flee from; pleasure is what we cling to and labour to increase'.[60] In controlling the amounts of pleasure and pain the valley-dwellers experience, Hinton's Persian king works on the assumptions of both Bain and Bentham. Unaware of the king's presence and feeling only the effects of the king's manipulations, the valley-dwellers construct a utilitarian understanding of themselves and their world. According to the narrator, 'the inhabitants knew that they sought pleasure and avoided pain, and the great object was to make their life more pleasurable'; they also knew that 'sensation was passing off into a form from which it never reappeared [...], hence they concluded that sensation in the valley was gradually running down'.[61]

Misunderstanding the cause of all their actions, and thus their survival, the valley-dwellers fear that when all sensation has been lost they will fall into apathy which, in their world, inevitably leads to death. Hinton never explicitly mentioned the fourth dimension in 'The Persian King'; rather, he used commonly accepted nineteenth-century scientific theories to construct an allegory that is actually a subtle critique of the epistemology out of which those very theories arose. Here Hinton pushed through the second law of thermodynamics, inverting the implications of dissipation. The pain-absorbing Persian king is the 'permission' that allows for all life and development in the valley rather than 'the gradual annihilation of life', which is how the scientists of the valley interpret his actions.[62]

At the time in which the main narrative is set, the civilization of the valley is a secular, scientific society. The plot begins to take shape when a university student in the valley's metropolis is exiled to the outer, rural regions of the civilization as punishment for questioning one of the fundamental physical laws of the valley. The student encounters the folk beliefs of the rural agricultural workers, who are still somewhat superstitious and acknowledge the existence of the Persian king. However, the rural valley-dwellers also misinterpret the king's role in that they understand him to be an omniscient presence that is pained by any pleasure experienced by the inhabitants of the valley: 'They thought it pained him when they had pleasure, but not in the way in which was really the case. They thought simply that it was pain to him to see them taking pleasure'.[63] The student is able

---

59 Bain, *The Senses*, 325.
60 Ibid., 89.
61 Hinton, 'The Persian King', *Scientific Romances*, 64 and 70–71.
62 Ibid., 71.
63 Ibid., 78.

to apply his scientific education to the mythology of the rural valley-dwellers in order to discover the truth of the king:

> Now the student saw clearly some errors, some contradictions in their belief. For instance, he knew that beings only followed pleasure, and directly pleasure was equalled by pain, sank into apathy, and then gradually vanished away. Hence, he knew there need be no apprehension of the power's acting as they thought. He did not approve of the results in their life, for it was in consequence very gloomily framed [...]. But he knew as a scientific fact that there was constant diminution of feeling; and since he also knew that beings in the valley did nothing except it was more pleasant, he concluded that although pleasure and pain might both be disappearing, still pain must be disappearing to a greater extent. Now since the feeling did not become nothing, but passed away out of the perception of the inhabitants, it followed that it must pass away to some being. It did not disappear as feeling, but passed away from the sensation of the inhabitants. Is there a being, then, he asked himself – the power of whom these simple folks tell – who bears the difference of pain, and so makes existence pleasant to us? And is that the meaning of what they say that our pleasure pains him? Is it just the truth read backwards[?][64]

Here Hinton began to challenge the conflation of scientific epistemology with religious ontology that resulted in the moralization of the second law of thermodynamics. The student is able to work out the actual meaning of the king's action and thus transcend the pleasure/pain dynamics that underpin the valley-dwellers' utilitarian understanding of themselves and their world. To understand pain not as a motivating psychological force, but rather as an epistemological effect, is to recognize the limited nature of scientific epistemology in the valley, just as the valley's scientists recognize the limits of rural religious folklore.

The dogmatic, and thus limited, nature of both science and religion is a theme that recurs throughout Hinton's work.[65] The very title of his *Scientific Romances* is significant for its combination of two opposing discourses: science, or realism, which prioritizes objective observation and empirical evidence, and romance, or fantasy, which celebrates intuition and the imaginative capabilities of the individual subject. Here, again, we see the influence of James Hinton's lawbreaker: Hinton, like the student, challenged the second law of thermodynamics and, by extension, the theological implications attached to it. To some extent the pain-bearing king also demonstrates James Hinton's notion of service; certainly the conclusion that personal pleasure is not inherently morally suspect was influenced by James Hinton's philosophy.

Earlier in the history of the valley, before the student discovered the truth, the Persian king had revealed himself and his work to another valley-dweller, a prince. The king made contact with the prince, Hinton explains, because he was lonely. The king selected the prince because he, too, was 'destined to reign in his turn over a numerous people'.[66]

---

64 Ibid., 78–79.
65 In his final text, which was in press at the time of his death, *An Episode of Flatland*, Hinton's autobiographical character, Hugh Farmer, rails against the dogmatism of theology, continuing on to remark that 'the dogmatism of scientific men is stronger than the dogmatism of religion' (73–74).
66 Hinton, 'The Persian King', *Scientific Romances*, 52.

To further increase the prince's understanding, the king bestowed upon him the power of bearing pain of the other valley-dwellers. However, the king's revelation to the prince had disastrous results. The prince was unable to cope with his newfound knowledge:

> 'One thing succeeds another in the valley; pain follows pleasure, and pleasure follows pain. But the cause of all being is in bearing pain. Wherefore,' he cried, 'let us seek an end to this show. Let us pray to be delivered, that at last, pain ceasing, we may pass into nothingness.'[67]

The student reacts very differently to his discovery; he does not 'look upon nothingness as the desired end of existence. He felt the presence of one he discerned through thought, and this seemed more real to him than life or death'.[68]

The difference in the reactions of the prince and the student result from the difference in the ways in which they learn of the king: the king reveals his powers to the passive prince, while the student must undergo a process of learning unassisted by the silent and unseen king. After living with the rural valley-dwellers and learning of their beliefs, the student studies a number of historical and scientific texts, undergoing a period of speculation and introspection that leads to his discovery of the king. Having come to the conclusion that the rural valley-dwellers have simply read the truth 'backwards', the student goes out walking alone one night. Unlike the prince, the student does not receive any direct communication from the king:

> Now it may be considered surprising that the king did not communicate in some way with the student, for by means of his rays he was in possession of all that had gone on in his mind. But the king had found over and over again that if he manifested himself to any one of the inhabitants of the valley, the effect, though good at the immediate time, was most disastrous for the following time. [...] So when the student went out into the open air he saw nothing except the stars, and heard nothing except the wind. [...] He had not gone far when he saw a kind of luminousness. Is the moon beginning to rise? He thought. But he found he had passed the light and was leaving it behind. He could not have passed the moon thus. He went towards the light, and when he had reached it, it seemed like a slender staff of light.[69]

This staff of light allows the student to share in the king's pain-bearing ability, as the student quickly discovers through a process of trial and error. The king neither appears nor speaks to the student at any time. Unlike the prince who acquires the knowledge of the king by revelation – a 'divine right' of sorts – the student must learn and earn for himself knowledge of the king; it is the process of discovery that prepares him for this knowledge.[70]

---

67 Ibid., 53.
68 Ibid., 98.
69 Ibid., 80.
70 The differences between the prince and student are also legible as analogous to an opposition of Judaism to Christianity that privileges the latter. The prince, like Moses, is the passive recipient of the king's revelation, while the student must work it out for himself, as an 'everyman'. In

The student is therefore able to fully comprehend the king's role in the valley: he does not conflate the king's action of absorbing pain with the notion of an omnipotent 'first cause'. In his attempts to explain the nature of the king's action to a friend, the student observes that 'a cloak has been woven. The nature of [the king] is hidden. His nature has been connected with introspective questions about the origin – of, of all things, the way in which we perceive'.[71] Here we return to Hinton's larger project of accessing the fourth dimension. The Persian king is an allegorical representation of the fourth dimension, that – as in Stewart and Tait's unseen universe – absorbs the energy lost to the visible universe. The student must learn of the fourth dimension without divine assistance; in feeling 'the presence of one he discerned through thought', the student finds something 'more real to him than life or death'. By making what Hinton's contemporary, John Tyndall, described as 'a leap of the prepared imagination', the student is able to discover 'a world not less real than that of the sense, and of which the world of sense itself is the suggestion and justification'.[72]

Because he occupies the world behind the senses, the king has become associated with the omnipotent and the supernatural, but he is neither of these. The king is limited, and the fact that the valley-dwellers refuse to accept the student's discovery and eventually execute him for his radical beliefs seems to be a reaction informed by the earlier religious teachings in the valley. The notion that such a being exists and that he condones the suffering of the inhabitants of the valley – or worse, yet, is incapable of preventing suffering – is possibly more disturbing than the presumed non-existence of the king. Such is the view of the student's friend: 'This seems to me a very dismal doctrine. I can imagine some poetry in the idea of a being of infinite power, strong and glorious, but none in the idea of a suffering being'.[73] The sublime idea of infinite power, the 'poetry' of an omnipotent king, is the pathetic fallacy that lies at the root of the problematic nature of epistemology within the valley, as the student observes:

> Whatever we apprehend, we apprehend as powerful. Now since this quality of powerful comes in with regard to everything, it is probably introduced by the mind, and is rather a part of the mental action in giving an idea of reality than a quality of reality. [...] Of course, if we think of [the king] at all, we must conceive of him as powerful; *the nature of our mental action demands this.*[74]

---

this reading, the prince's failure to discover self-determination through a kind of Nietzschean will-to-power (or, perhaps, more appropriately a 'will-to-passivity'), parallels Moses's failure to enter the Promised Land. Like Jesus, the student is denied the king's assistance, even when he is martyred. Unlike Jesus, however, the martyring of the student does not result in salvation for the valley-dwellers; rather, their refusal to acknowledge his message of self-determination results in the collapse of their civilization.

71 Ibid., 85.
72 Tyndall, *Fragments*, 128 and 132.
73 Hinton, 'The Persian King', *Scientific Romances*, 83.
74 Ibid., 85, emphasis added.

Instructive here is Ernst Cassirer's concept of 'momentary gods' created

> when external reality is not merely viewed and contemplated, but overcomes a man in sheer immediacy [...] then the spark jumps somehow across, the tension finds release, as the subjective excitement becomes objectified, and confronts the mind as a god or daemon.[75]

Drawing on Victorian philologist Max Müller, Cassirer argued that these 'momentary gods' are at the root of human language and epistemology. Hinton viewed this as the limited 'nature of our mental action' and wanted to push beyond it. In overcoming the pathetic fallacy with reference to the action of the invisible king, the student of 'The Persian King' surpasses Ruskin's highest order of poet. He has become a sort of hyper-perceiver, one who has faced forces 'inconceivably above' him and not seen 'untruly'.[76] The king is no god for the student, momentary or otherwise. However, the king – as a representative of the fourth dimension in this story – shares affinity with the demon of Maxwell's thought experiment.

Maxwell's proposition highlighted what Clarke calls the 'crucial if debatable observation that entropy is [...] an epistemological effect, a product of the limitations of human perception[,] for instance, our inability to manipulate matter at the molecular level'.[77] Hinton was also concerned with exposing the limitations of both scientific and religious epistemologies, with their shared emphases on origins and endings. In the direct discourse that forms the second part of 'The Persian King', Hinton unpacked his allegory, explaining that in the physical sciences, 'we have thought of motion as a thing in itself impaired by the multitudinous obstacles it meets in the world'.[78] The result of interaction between objects is friction; through friction, a small portion of heat that is generated is dissipated, never again to be accessible. But rather than stop here, where the science of thermodynamics – limited as it is by the three-dimensional, linear temporal logic of both science and theology – does 'let us look on the circumstances more impartially. Let us look on them as something co-equal with motion. Let us find in that mode whereby all motion comes to an end[,] the originating cause also whereby all motion comes to be'.[79]

The king animates the valley by a paradoxical act of will-to-passivity; he is 'bearing rather than exerting force'.[80] The force itself does not originate with the king, who actually functions as an absence, a void that allows sensation to 'pass off'. As the king bears a portion of the painful sensation of all actions for the valley-dwellers – creating an imbalance in sensation that initiates a prompt to action – so does the

---

75 Cassirer, *Language and Myth*, 33.
76 Ruskin, *Works*, 5: 209.
77 Clarke, *Energy Forms*, 119.
78 Hinton, 'The Persian King', *Scientific Romances*, 108.
79 Ibid., 108.
80 Ibid., 127.

disappearance of heat in the three-dimensional universe allow for movement. Hinton explained:

> The ultimate transformation of all energy of motion is into the form of heat. [...] This passing of energy into the form of heat must not be regarded as a side circumstance, as less essential to the laws of nature than that law we call the conservation of energy.[81]

In Hinton's interpretation, what is perceived by most as deadly chaos is actually a positive, creative force. This void is constitutive, similar to the transvaluation of negative space that Stephen Kern identifies as characteristic of the early modernist aesthetic:

> The traditional view that space was an inert void in which objects existed gave way to a new view of it as active and full. [...] I will refer to this new conception as 'positive negative space.' Art critics describe the subject of a painting as positive space and the background itself as negative space. 'Positive negative space' implies that the background itself is a positive element, of equal importance with all others.[82]

This shift in perception is indicative of the beginnings of a trend of acceptance and even celebration of disorder and the relativity of perception and, as Clarke observes: 'This unusual championing of dissipative processes – an appreciation for, rather than denigration of, friction and resistance – is the truly predictive portion of Hinton's text'.[83]

Hinton's treatment of causation is equally indicative of the shift away from the model of reliance on a singular deity possessing 'infinite energy', as we see in Stewart and Tait's theological cosmology. 'In past times', according to Hinton,

> people really felt sure about certain things being causes which we now know had a very slight connection with the result. Incantations have been supposed to have an effect on physical phenomena, such as eclipses. [...] To say one external event is the cause of another is to put an absolutely unknown and spiritual relation in the place of impartial observations. [...] To be the antecedent in a chain of movements is the fact which we can observe about any movement in the external world. We cannot strictly say what movements of gases, water, &c. cause this volcano. We can only say what movements of gases, water &c., precede this volcanic eruption analogous to movements which have preceded other volcanoes.[84]

Physical events in the external world do not function as causes in themselves, Hinton argued. When writing of a 'true' causal relationship, he described it as a 'spiritual and unknown relation': 'To cause a motion is the name for the action of our soul upon matter'.[85] Here causation involves an act of human will: '*We* are the cause of actions we will. The notion of a cause is derived from our "will" action, and the notion of cause ought to

---

81 Ibid., 107.
82 Kern, *The Culture of Time and Space*, 153.
83 Clarke, *Energy Forms*, 120.
84 Hinton, 'The Persian King', 108–10.
85 Ibid., 110.

be kept to this connection'.[86] Therefore, when observing a chain of events in the external world, it is impossible to attribute a 'theory of mind' to any particular event.[87] However, omitting causation 'from the external chain of events' does not necessarily preclude the possibility of any agency in the external world, according to Hinton: 'Let us not introduce the notion of causation at haphazard. But if we find in the external world signs of an action like our own will action, let us then say, Here is causation'.[88] To be a source of 'true' causation – in the context of Hinton's greater project – is to be a 'four-dimensional agency'.

Nevertheless, this four-dimensional agency should not be read as necessarily implying a divine first cause. Hinton had already suggested that human beings are actually four-dimensional agents in 'What Is the Fourth Dimension?'. As I observed in the previous chapter, there is an avoidance of discussion of origins in much writing about the fourth dimension, Hinton's included. While Hinton often described the fourth dimension as if it were *the* transcendent metaspace, throughout his work he simultaneously undermined any attempt to read hyperspace philosophy as an absolutist project. This tension is observable throughout Hinton's work, and in 'The Persian King' we can observe the open-endedness of Hinton's project in the relationship between the king and Demiourgos. This relationship is collaborative, as is, by analogy, the relationship between Hinton and the reader. Although it is through the agency of the king that the valley-dwellers are animated, he is not the cause of their existence. The king merely sets the beings into motion by following Demiourgos's instructions; Demiourgos has created them by playing music on his pipe. Demiourgos's powers are similarly limited; he is neither able to animate the valley-dwellers himself, nor can he create them so that they will be self-animating. Therefore, we cannot observe a singular 'first cause' within this narrative, and it is here again that we encounter the limits of language. A first cause cannot be explained by analogy because nothing corresponds to it. Thus, as Cassirer observed, Yahweh of the *Old Testament* can only explain himself to Moses as 'I am that I am'.[89] This is pure tautology, and can be extended to a meaningless infinite series.

According to Hinton, the only way to make such an infinite series cognizable is to impose an artificial limit on it, usually through the device of personification. Hinton acknowledged the false limitations he sets up in 'The Persian King', writing that, when using the king as a personification of 'an ultimate medium', 'it must be remembered that this conception of an ultimate medium was merely a supposition to enable us to see and roughly map out the relations of the things we were investigating. Where we were really landed was an infinite series'.[90] According to Hinton, an infinite series typically appears as a result of using flawed instruments of measurement or observation. Hinton turned to algebra for an example: 'Infinite series occur when the object which it is wanted to represent in algebraical terms cannot be grasped by the algebra'. Just as algebra breaks down

---

86 Ibid., 108, emphasis added.
87 See Zunshine, *Why We Read Fiction*.
88 Hinton, 'The Persian King', *Scientific Romances*, 110–11.
89 Cassirer, *Language and Myth*, 77.
90 Hinton, 'The Persian King', *Scientific Romances*, 120.

when it tries to represent the 'trigonometrical idea' of *cosine x*, language, which is based on the presently limited perceptual abilities of the three-dimensional human consciousness, begins to break down when it encounters the fourth dimension. In algebra, 'when there is no single term or set [...] which will serve, the object is represented by means of an infinite series'.[91] Thus, working with a similarly limited instrument (language) Hinton provided the reader with a series – multiple though not infinite – of texts, each of which presents a different perspective on the fourth dimension.

The personification of the king is used in the first part of 'The Persian King' because it is difficult to describe or explain a void or absence in the place of a divine agent. This difficulty arises out of one of the limitations of language which, as Beer has observed, 'is anthropocentric. It places man [*sic*] at the centre of signification'.[92] Thus, there are no 'terms' that can grasp such a concept as Hinton's void in the place of a divine agent, just as in 'What Is the Fourth Dimension?' he struggles to describe the appearance of a four-dimensional object. Hinton's hyperspace philosophy paradoxically supports and undermines humanism. On the one hand Hinton is like Derrida's *bricoleur* in that his hyperspace philosophy is 'no longer turned toward the origin', and his methodology in the *Scientific Romances*

> affirms freeplay and tries to pass beyond man and humanism, the name man being the name of that being who [...] through the history of all his history – has dreamed of the full presence, the reassuring foundation, the origin and the end of the game.[93]

On the other hand, Hinton argued that human beings are capable of transcending their current limitations and realizing their hyperbeing potential as four-dimensional agents.

Hinton described those who possess this four-dimensional consciousness as 'true personalities, conscious of being true selves, the oneness of them lying in the [fourth dimension], but each spontaneous in himself and absolute will, not to be merged in any other'.[94] The student in 'The Persian King' attains this status and, like the king, is able to absorb the pain of others through a kind of will-to-passivity. Thus, the student becomes a true agent of causation; it is because of this that the politicians and scholars of the valley find him threatening and repugnant. 'He made me feel like a puppet', one council member complains, and even the student's friend admits that 'he seems to lack the ordinary springs of motive'. Fearing his influence on the community, that they too may become 'difficult to govern' if they transcend their mechanical existence of pursuing pleasure and avoiding pain, the council has the student executed.[95]

After the student's death, the king departs from the valley because, he explains, he is 'weary'. The fate of the valley-dwellers is the same as that proposed by Stewart, Tait and other proponents of cosmic heat death:

---

91 Ibid., 121.
92 Beer, *Darwin's Plots*, 47.
93 Derrida, 'Structure', 254.
94 Hinton, 'The Persian King', *Scientific Romances*, 128.
95 Ibid., 93–94.

As soon as the king had departed from the valley the beings in it began to sink into the same state of apathy as those were whom he had first found there. […] A chill death in life crept over the land. 'Tis useless to ask after the fate of any one of those that were there, for each was involved in the same calamity that overwhelmed all. […] The busy hum of life in the streets was hushed. […] In every spot was such unbroken quiet as might have been had all the inhabitants gone to some great festival. But here was no return of life. No watchful eye, no ready hand was there to stay the slight but constant inroads of ruin and decay. The roads became choked with grass, the earth encroached on the buildings, till in the slow consuming course of time all was buried – houses, fields, and cities vanished, till at length no trace was left of aught that had been there.[96]

In refusing to acknowledge the limited nature of their current lives, the council members of the valley damn themselves to a universal death. For Hinton, the struggle to transcend the three-dimensional is not the same as Stewart and Tait's redemption through 'a universe possessing infinite energy', with a 'developing agency [which] possesses infinite energy'.[97] Rather, the reader – like Hinton's student – is encouraged to use both 'modes of access' to the king/fourth dimension: one is through empirical knowledge of the outside world, or scientific understanding, and the other is through the self, or Romantic introspection. To succeed in this endeavour is to become a kind of hyperbeing, a true personality, unable to be influenced by god or human.

'Casting Out the Self'

The final text within the first series of the *Scientific Romances* provides Hinton's first attempt to access the fourth dimension through practical exercises with model cubes. Here the reader is guided through the manipulation of 27 cubes, whereby, through an act of aesthetic will, they are to 'cast out the self'. Again we can see the oppositional forces within Hinton's hyperspace philosophy: the stated purpose of the cube exercise in 'Casting Out the Self' is to strip the thinking subject of its subjectivity, to break down the boundary between self and other. The problem is that to actually accomplish such a task would involve annihilation of the self, to become Ruskin's paradoxical invisible man. Thus, Hinton's project partakes of the 'suicidal narrative of knowledge' that George Levine identifies as particularly resonant in nineteenth-century scientific culture.[98] In 'Casting Out the Self' Hinton proposed a radical deconstruction of the self while at the same time anxiously guarding against total dissolution by maintaining agency through emphasis on the manipulation of physical objects. Here Hinton was attempting to expel what he saw as a false limitation on human consciousness: the three-dimensional self.

What Hinton viewed as an extractive ('casting out'), deconstructive activity in this text actually becomes a creative process; the manipulation of cubes in 'Casting Out the Self' is analogous to the reading process that is engendered by the genre variation within

---

96 Ibid., 101.
97 Stewart and Tait, *The Unseen Universe*, 220.
98 Levine, *Dying*, 5.

Scientific Romances. In his hyperspace philosophy Hinton seemed to be simultaneously questioning and guarding against the unexpected and accidental. As Bell has noted, to question is to foreground the liminal nature of subjectivity. By stripping away the three-dimensional self, Hinton sought to construct a daemonic four-dimensional subject that necessarily exists at the boundary of things.

'Casting Out the Self' is the most explicitly collaborative of all the texts in the first series: by instructing the reader through a number of exercises with a set of model cubes, Hinton sought to change the reader's intuition of space from an unconscious activity to a conscious one. To achieve this, it is necessary first to become self-conscious, and then to 'cast out' all elements of self. At work here is what Levine identifies as 'a view [that is] built into the idea that the senses that are the gateway to all knowledge must be disciplined and checked in order to provide that knowledge'.[99] Hinton called on the reader to undergo a process, the function of which is to deny the subject's normal, three-dimensional sensory perceptions in order to uncover what he claimed was a more direct, unmediated encounter with 'true' knowledge: intuition of the fourth dimension.

While in many ways Hinton's hyperspace philosophy is directly opposed to Henri Bergson's *la durée pure*, Bergson's definition of intuition in his *Introduction to Metaphysics* (1903) is particularly apt here: 'By intuition is meant the kind of intellectual sympathy by which one places oneself within an object in order to coincide with what is unique in it and consequently inexpressible'.[100] Hinton wanted the reader to physically and psychically identify with each cube, and in this way to 'feel' space. In this final romance of the first series, Hinton attempted to transcend analytical conception and perception of the fourth dimension, to gain an intuitive knowledge of space that would allow for the four-dimensional imagination to develop. Again, Bergson is useful here:

> To analyse [...] is to express a thing as a function of something other than itself. All analysis is thus a translation, a development into symbols, a representation taken from successive points of view from which we note as many resemblances as possible between the new object we are studying and others which we believe we know already.[101]

To intuit, by contrast, is to know the thing from the inside out. To intuit is to no longer need to view the 'slices' of the fourth dimension successively; rather, the intuitive encounter with four-dimensional space allows the subject to surpass the Ruskinian imagination, to see all of the three-dimensional world simultaneously.

Hinton's first set of cube exercises in 'Casting Out the Self' appears to have grown out of a personal crisis of knowledge: 'The beginning of it was this. I gradually came to find that I had no knowledge worth calling by that name, and that I had never thoroughly understood anything which I had heard'.[102] Hinton's crisis of knowledge seems to be

---

99 Ibid., 2.
100 Bergson, *Introduction*, 7.
101 Ibid., 6–7.
102 C. H. Hinton, 'Casting Out', *Scientific Romances*, 205.

linked to a fear of self-dissolution and to a growing sense of mortality. The timing of this crisis is telling: in the text, Hinton describes it as occurring when he finished his formal education. 'Casting Out the Self' was originally published in 1886, the year he received his MA. Hinton was already living away from Oxford, working at Uppingham College as a mathematics instructor.[103] It was also the year of his bigamy conviction. However, 'Casting Out the Self' was composed before these events, and so, while the strain leading up to these various upheavals was no doubt building, the period of crisis he refers to is more likely to have occurred in the previous decade when another series of major life events occurred: in 1875, Hinton's father died; in 1876, he completed his undergraduate work at Oxford; within the year he had taken a post at Cheltenham Ladies' College as a mathematics instructor, and in 1880 he married his first and legal wife, Mary Ellen Boole. Of the timing of this existential crisis, Hinton wrote:

> I will not go into the matter further; simply this was what I found [that he knew nothing], and at a time when I had finished the years set apart from acquiring knowledge, and was far removed from contact with learned men. I could not take up my education again, but although I regretted my lost opportunities I determined to know something. [...] And I would earnestly urge all students to make haste in acquiring real knowledge while they are in the way with those that can impart it; and not rush on too quickly, thinking that they can get knowledge afterwards. For out in the world knowledge is hard to find.[104]

The sentiments expressed here are not atypical. Lacking, for the first time in his life, the structure provided by an institution of formal education, the narrator appears to be struggling against a sense of self-dissolution, figured here as the sudden dissipation of the illusion of knowledge. By undermining both his previously acquired knowledge and his ability to obtain knowledge in the future, the Cartesian model of the thinking subject is thrown into radical doubt. He is also anxious about facing the challenge of unstructured learning: 'Out in the world knowledge is hard to find'.

Hinton found comfort by creating another, highly structured, process of learning to undertake. This bizarre form of self-affirmation through self-abnegation resulted in his cube exercises: he began by memorizing the relative positions of 216 wooden cubes arranged in such a way as to compose one larger cube. For practical reasons, in 'Casting Out the Self' Hinton reduced the number of cubes needed to perform this exercise to 27, arranged in a 3 × 3 × 3 unit. 'Now', he wrote,

> this is the bit of knowledge on which I propose to demonstrate the process of casting out the self. It is not a high form of knowledge, but it is a bit of knowledge with as little ignorance in it as we can have; and just as it is permitted a worm or reptile to live and breathe, so on this rudimentary form of knowledge may we be able to demonstrate the functions of the mind.[105]

---

103 Ballard, 'The Life and Thought', 29.
104 Hinton, 'Casting Out', *Scientific Romances*, 206.
105 Ibid., 208.

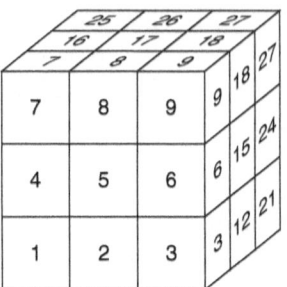

Figure 2.1

Hinton could just as easily have written that 'this is the bit of knowledge' on which he was able to demonstrate the existence of his subjectivity. By enabling himself to 'demonstrate the functions' of the mind, Hinton confirms the existence and the functioning of *his* mind. In this text, he represents knowledge – 'rudimentary' though it may be – as being indicative of a functioning mind, which he compares to the basic activities that are necessary to life in evolutionarily 'lower' species. According to Hinton, this 'bit of knowledge' is the knowledge of space relations. The manipulation of the cubes allows the subject to demonstrate its knowledge of the relations of the blocks to each other while simultaneously attempting to avoid any reference to the subject's position to the block of cubes as an external reference point, or as an agent.

For example, Hinton explained, 'if I say cubes 1 and 2, I mean the two which lie next to each other' (Figure 2.1).[106] The relation between these two cubes is the common side between them. Hinton memorized the relations between all 27 of the cubes, and then began the process of 'casting out the self'. 'First of all,' he explained, 'when I had learnt the cubes, I found that I invariably associated some with the idea of being above others. When two names were said, I had the idea of a direction of up and down.'[107] The idea of up and down are what Hinton calls 'self elements': 'I only conceive of an up and down in virtue of being on the earth's surface, and because of the frequent experience of weight'. This is a description of the experience of embodiment, what Hinton described as a 'condition affecting myself'.[108] The language here assumes a Cartesian dualism of mind and body, but Hinton's cube exercise is an attempt to access a psychical, four-dimensional self that has been incorrectly fused with a three-dimensional body. While underlying Hinton's view is the assumption of a transcendent, absolute reality, there is also space here for the radical enumeration of perspectives: the medium of the three-dimensional body causes the human mind to misapprehend one level of dimensionality as an exclusively singular reality. There is an awareness here that human perception of reality is not absolute; perception is affected by physical factors ranging from simple differences, such

---

106 See Fig. 2.1, Hinton, 'Casting Out', *Scientific Romances*, 208.
107 Ibid., 208–9.
108 Ibid., 209.

as variations in eye placement across species to the intricacies of neurological structure. With this in mind, Hayles explains:

> Our so-called observables are permeated at every level by assumptions located specifically in how humans process information from their environments [...] it becomes clear that observables really mean observations made by humans [...]. In short, we are always already within the theater of representation.[109]

Hinton appeared to regard his cube exercises as a means of stepping outside this theatre, a problematic project, as this theatre – physical embodiment – is also the framework that enables the human subject to exist. Hinton was not consciously seeking self-annihilation in the form of physical death; he did not seem to make this connection here. In his view it is only by stepping out of the theatre of three-dimensional embodiment that one can access the fourth dimension through the imagination. William J. Scheick notes that cubists – or at least their followers – similarly sought to encounter the fourth dimension by privileging imagination over sensory perception, writing that contemporary art critic Maurice Raynal 'concluded that Cubist artists conveyed this [fourth] dimension by "painting objects as they *thought* them" rather than how their senses perceived them'.[110]

Hinton decided that in order to fully escape the theatre of representation and access the realm of pure abstraction, one must 'cast out' the effects of gravity and 'handedness', perceptions of up and down and left and right.[111] The only way to remove these self-elements from one's knowledge of the block of cubes, Hinton claimed, was to turn the block upside down, invert it, and relearn the blocks in each new position:

> It was, I found, quite necessary to learn them all over again, for, if not, I found that I simply went over them mentally the way first learnt, and then about any particular one made the alteration required, by a rule. Unless they were learnt all over again the new knowledge of them was a mere external and simulated affair, and the up and down would be cast out in name, but not in reality. It would be a curious kind of knowing, indeed, if one had to reflect what one knew and then, to get the facts, say the opposite.[112]

Like the Persian king with his subjects, Hinton favoured experiential learning rather than memorization of a system of rules. However, his system for manipulating the cubes is carefully circumscribed within its own set of rules.

Following the 'correct' procedure for his cube 'games' requires, in Hinton's words, 'considerable mental effort' and time. For Hinton, however, it is the only means of approaching a way of knowing and seeing that is not limited by three-dimensionality.[113]

---

109 Hayles, 'Constrained Constructivism', 28.
110 Scheick, *The Splintering Frame*, 22, original emphasis.
111 'Handedness' was, for Kant, and for many present-day philosophers, a contentious issue concerning the opposition of absolutist and relationist concepts of space. See Le Poidevin, *Travels in Four Dimensions*, 64–66.
112 Hinton, 'Casting Out', *Scientific Romances*, 209.
113 Ibid., 210.

The reward, according to Hinton, is a radically new way of encountering reality. This new spatial awareness is a deconstruction of the 'I–thou' opposition: there is no up or down, left or right in the cube exercises, and thus, it is implied, there is no physical subject. In order to achieve this dissolution of the self, the exercises must be done physically: the reader must actually handle the blocks, not just conceive of them mentally. Thus, 'Casting Out the Self' is a text that resists mere consumption. As a Text, this romance requires that the reader '*set it going*'.[114] Once such textual game-playing has begun, according to Iser:

> Author and reader are able to share the game of the *imagination*, and, indeed, the game will not work if the text sets out to be anything more than a set of governing rules. The reader's enjoyment begins when he himself [*sic*] becomes productive, i.e., when the text allows him to bring his own faculties into play.[115]

While Hinton certainly invited his readers to share in this 'game of the imagination' that is intuiting the fourth dimension, this creative impulse sits somewhat uneasily with the stated purpose of 'Casting Out the Self', to remove most of the reader's faculties from play. The very ineffability of the spatial fourth dimension makes Hinton's project a highly creative and unstable one: no matter how carefully he laid out 'rules' and guidelines for the mental construction of the fourth dimension, this act of construction is necessarily the responsibility of the reader. Thus, there are as many fourth dimensions as there are readers of – or participants in – Hinton's texts.

The maintenance of an unstable unity is built into the structure of the *Scientific Romances*: each individual text in this collection plays off the others, mimicking Hinton's hypothesis that the fourth dimension can only be perceived from the three-dimensional perspective as a series of 'slices'. Moving from one 'slice' to another – the perspective engendered by the 'instructions' contained within each individual text – the reader experiences what Iser described as the 'concrete fluidity' of the text, where 'the reader is constantly feeding back reactions' as they encounter 'new information'.[116] In a manner echoing Hinton's 'Arabic method of description', each succeeding viewpoint of the fourth dimension offered to the reader within the *Scientific Romances* builds upon the previous ones, at the same time illustrating that, in Iser's words, 'the whole text can never be perceived at any one time'.[117] It was Hinton's goal, however, to instil in his readers a simultaneous and unifying aesthetic capability that transcends Ruskin's highest human imaginative achievement. Although Hinton often wrote of the fourth dimension as a transcendental metaspace, the dynamism of his *Scientific Romances* – the insistence on obtaining knowledge through relations and movement of consciousness – resists an absolutist, monological model of reality. The game can only be played – the Text can only instruct the reader on intuiting the fourth dimension of space – by offering itself 'to a

---

114 Barthes, 'From Work', 197, original emphasis.
115 Iser, *The Act of Reading*, 108, emphasis added.
116 Ibid., 68.
117 Ibid., 108.

diffraction of meanings' and the dynamism of his *Scientific Romances*, with its insistence on obtaining knowledge through relations and movement of consciousness, would appear to belie this totalizing goal.[118] Thus we observe the paradoxical nature of Hinton's hyperspace that he insisted is a transcendent, yet material, realm that the unified, embodied subject can only encounter through diffracted, multiple 'approaches'.

Referring to his own trial-and-error experimentation with developing intuition of the fourth dimension through his cube exercises, Hinton claimed in *A New Era of Thought* that 'I can lay it down as a verifiable fact, that by taking the proper steps, we can feel our four-dimensional existence, that the human being somehow, and in some way, is not simply a three-dimensional being – in what way it is the province of science to discover'.[119] Hinton was careful to remind the reader of the practical implications of his hyperspace philosophy: 'We have to choose between metaphysics and space thought', he explained, because while 'in metaphysics we find lofty ideals', metaphysics without practical application 'reduces the world to a phantom and ourselves to lofty spectators'.[120] Unlike metaphysics, Hinton's more practical 'space thought' must be approached through active engagement with the world. By following his instructions, he explained, we can learn to intuit the relations within a block of cubes without reference to the human body, and then

> we discover in our own minds the faculty of appreciating the facts of position independent of gravity and its influence upon us [...]. The discovery of this capacity is like the discovery of a love of justice in the being who has forced himself to act justly. It is a capacity for being able to take a view independent of the conditions under which he is placed, and to feel in accordance with that view.[121]

Here Hinton claimed asceticism as the means by which the four-dimensional aesthetic can be achieved. He implied that the view independent of one's surroundings – and the view that accompanies this independence – is the perspective accessible from the fourth dimension. This is a perspective that is not available to those who are simply 'lofty spectators', but only to those 'who ha[ve] forced' themselves 'to act justly'. Knowledge and vision are necessarily linked to morality here: to 'see' from this perspective is to 'feel in accordance with that view', and thus have a personal stake in it.

The paradox of aesthesis through ascesis is part of what Levine calls the 'dying to know' narrative that underpins nineteenth-century scientific epistemology and, perhaps surprisingly, aesthetics as well:

> Ascesis, though almost always taken as a condition for the disinterest and objectivity necessary for adequate scientific observation and experiment, is also a way beyond the entrapment of the self, both a social and aesthetic extension of the particulars of individual sensation to the shareable conditions of true knowledge. [...] When Pater insists on the necessity 'to know

---

118 Barthes, 'From Work', 195.
119 Hinton, *A New Era*, 46.
120 Ibid., 37.
121 Ibid., 32–33.

one's own impression as it really is, to discriminate it, to realise it distinctly,' he is pointing to a truly ascetic activity that is often mistaken for self-indulgent. Experiencing sensation is one thing; literally knowing it, discriminating it, entails repression of self, almost a denial of the sensation being experienced. Pater is arguing for a way to be objective about subjectivity, to find a position outside experience from which to experience it and possess it.[122]

Likewise, for Hinton, to know an object from the inside is to feel as it 'feels', to be able to isolate 'self-elements' – or sense-impressions – and exclude them. To feel empathetically in accordance with something other than self is contingent upon a certain lack of egoism: hence, there is a desire to 'cast out' or dissolve the self. Just as Pater wanted to transcend experience in order to fully possess it, Hinton sought extension of self through dissolution: for example, Hinton claimed that the 'capacity for being able to take a view independent of the conditions under which' the embodied subject exists 'can only be done by, as it were, extending our own body so as to include certain cubes, and appreciating then the relation of the other cubes to those'.[123] If we read 'casting out the self' as extending the body – or the sensation of the embodied subjectivity – to encompass other, external perspectives, the title of this text takes on an entirely different meaning from that of self-surrender. Rather, here the self is 'cast out' like a net, in order to draw in objects around it. The self thus becomes a larger agent; like the student who gains access to the Persian king's light rays, the self here gains power through its violation of the dichotomy of internal and external.

After establishing a means for conceiving and perceiving the fourth dimension in the first series of *Scientific Romances*, in the second series Hinton turned his attention inward, to an exploration of the social, ethical and personal implications of his hyperspace philosophy for the individual subject. Although throughout this book I continue to revisit the ideas Hinton expressed in the first series, in the next chapter I shift my focus to the second series of *Scientific Romances*.

---

122 Levine, *Dying*, 249.
123 Hinton, 'Casting Out', *Scientific Romances*, 33.

Chapter Three

THE FOUR-DIMENSIONAL SELF: PERSONAL, POLITICAL AND UNTIMELY

Although I have noted that Hinton shifted to exploring the social, ethical and personal effects of his hyperspace philosophy in the second series of his *Scientific Romances*, I do not wish to imply a total lack of concern with the personal or political in the first series. In subverting the second law of thermodynamics in 'The Persian King', Hinton was not just providing a way out of cosmic heat death brought on by entropy. Bruce Clarke describes Hinton's treatment of the dissipation of heat as a reversal of the moral polarity of the second law of thermodynamics, which was typically associated with decadence and degeneracy.[1] While not all writers went as far as Stewart and Tait in admonishing the 'wasteful character' of the dissipation of heat, the term 'dissipation' itself carries with it connotations of moral condemnation.[2] M. Norton Wise observes that the principle of entropy expressed by Thomson in 'On a Universal Tendency in Nature to the Dissipation of Mechanical Energy' in fact 'reflect[s] the easy analogy between the degraded state of energy [...] and the degraded state of the laboring poor'.[3] Like late-Victorian anxieties about biological degeneration, the discourse of Victorian thermodynamics was often utilized by political and cultural conservatives to pathologize those who challenged traditional Victorian hierarchies. It is no coincidence that Stewart and Tait described 'the tendency of heat [...] toward equalisation' as *'par excellence* the communist of our universe'.[4] Just as the equalization of heat leads to a degraded state of energy, the degradation of humanity would result from the levelling of class distinctions.

Noting the gendered dichotomy of nature and culture in industrialized models of time management and productivity, Wise and Patricia Murphy observe how – in nineteenth-century capitalist discourse – linear, irreversible time and productivity came to be associated with more prestigious masculine work while repetitive, and 'reproductive' labour was gendered as feminine.[5] In industrialized urban areas, the labouring poor were seen to conglomerate in 'unruly crowds and mobs' that, in one Victorian statistician's words, 'substitute, for a population that accumulates and preserves instruction and is steadily progressive, a population that is young, inexperienced, ignorant, credulous, irritable,

---

1   See Clarke, *Energy Forms*, 111.
2   Stewart and Tait, *The Unseen Universe*, 197.
3   Wise, 'Time Discovered', 52.
4   Stewart and Tait, *The Unseen Universe*, 126, original emphasis.
5   See Wise, 'Time Discovered', and Murphy, *Time Is of the Essence*.

passionate, and dangerous, having a perpetual tendency to moral as well as physical deterioration'.[6] These 'morally degenerate' crowds are also effeminized, as Wise notes; they are either depicted as young, 'mere boys', or as a female, potentially revolutionary and bloodthirsty mob, as in Dickens's *A Tale of Two Cities*.[7] Hence, we can observe another specific layer of anxiety embedded within the discourse of Victorian entropy: the fear that eventually the balance would tip between the organized, 'productive' members of society and the chaotic, dangerous underlings. There is thus potential for a political reading of the inversion of entropy in 'The Persian King' that resonates with contemporary socialist and feminist movements.

In the present chapter I examine Hinton's second series of *Scientific Romances* (1896), a collection of texts that deals with the personal and political aspects of hyperspace philosophy in a more conscious manner. As with the last chapter, I draw on all the romances of the second series in my discussion, but my main focus is on two romances in particular, *Stella* and *An Unfinished Communication*. These two texts were first published within a single volume titled *Stella and An Unfinished Communication: Studies of the Unseen* (1895). This volume did not sell as well as expected so the two texts were reprinted, along with 'The Education of the Imagination' and 'Many Dimensions' as the second series of *Scientific Romances*.[8] Before examining these texts, it will be helpful to consider the political and social context of James Hinton's circle of influence in fin-de-siècle London.

Progressive thinkers such as Havelock Ellis, Olive Schreiner and Edward Carpenter were debating to what extent James Hinton's idea of 'service' might be useful in their attempts to rethink social and sexual relations. While the younger Hinton's *Stella* and *An Unfinished Communication* are both fascinating in their own right as literary texts, they also function as thought-experiments for Hinton to explore what role his theory of four-dimensional consciousness might play within late nineteenth-century progressive discourse.

Hinton acknowledged that the moral transformation that comes with the recognition of higher space can have radical and 'dangerous' consequences. The claim that 'self-regard is to be put on one side'

> is, in truth, a dangerous doctrine; at one sweep it puts away all absolute commandments, and all absolute verdicts about things, and leaves the agent to his own judgment. It is a kind of rule of life which requires most absolute openness.[9]

The language here echoes James Hinton's free-love philosophy of lawbreaking and service, and would have been familiar to his contemporaries as part of radical socialist and progressive sexual discourse. Hinton was aware of how his hyperspace philosophy

---

6   Wise, 'Time Discovered', 52.
7   Ibid., 52 and 47.
8   See Sonnenschein to Hinton, 10 March 1896, Letter Book 27, 956. All letters from Sonnenschein to Hinton are from the Publishers' Archives, MSS 382, 'Records of George Allen & Unwin, Ltd', Archive and Manuscript Division, Special Collections, University of Reading.
9   Hinton, *A New Era*, 90–91.

meshed with his father's ideas: for example, Hinton conflated his hyperspace philosophy and bigamy scandal in a letter to his publisher, William Sonnenschein. Referring to his manuscript for *A New Era of Thought*, Hinton explained that 'it is I believe the prosecution of this line of thought that the access of Science to those truths which are apprehended by the religious consciousness will be found'. Hinton noted that he had been struggling towards this realization himself; 'however [...] the effect has been thus far simply ruinous [...] I have found myself in a false position and the first and the absolute condition of any true life as I now understand it now lies in absolute openness'.[10] Hinton was alluding to his bigamy conviction here, noting that he was about to leave for a foreign appointment and that 'I have had to give up everything and go through disgrace [...] and have to put up with misconception on every side'. Sonnenschein avoided the subject of Hinton's scandal in his reply, writing,

> I fear I am too much of an ordinary-minded individual to fully enter into your thoughts. I consider your speculations, so far as I have examined them, of much interest; but it appears to me that their application to every-day practice is fraught with much risk of error, not to speak of so mean a thing as danger. [...] I have sent your book to press today [...] I was not aware that you had obtained a foreign appointment. What and where is it?[11]

The faint tone of disapproval within an otherwise uncharacteristically chatty reply from Sonnenschein points to his uncomfortable awareness of the scandal. Hinton's trial was briefly the talk of progressive circles in London, and Sonnenschein would no doubt have known about it. Not only did reports of the trial appear in national and local newspapers across England, but Sonnenschein was publisher for key members of the Fellowship of the New Life and the Fabian Society. James Hinton's followers were members of both societies, and gossip about the younger Hinton travelled fast.

The Hintonians

Before his death in 1875, James Hinton had already made a name for himself as a fashionable Harley Street aural surgeon.[12] However, if he is remembered for anything today, it is his writings in social philosophy and mysticism. Seth Koven has aptly described James Hinton as 'a philanthropic hedonist', and a number of his contemporaries embraced his philosophy because he argued that altruism need not be divorced from personal pleasure.[13] James Hinton remained a controversial and divisive figure for decades after his death; if anything, his teachings became even more controversial in the 1880s – for example, causing friction within the Havelock Ellis–Olive Schreiner–Karl Pearson triangle.[14]

---

10 Hinton to Sonnenschein, day and month illegible, 1887. Loose letter. Due to the biographical details, I estimate this letter was written in January or February of 1887.
11 Sonnenschein to Hinton, 25 February 1887. Letter Book 6, 834.
12 Originally an associate of Joseph Toynbee, James Hinton took over his practice after Toynbee's death in 1866.
13 Koven, *Slumming*, 16.
14 See, for example, Grosskurth, *Havelock Ellis*, and Nottingham, *The Pursuit of Serenity*.

Away from any hint of the scandals surrounding James Hinton's personal life, Havelock Ellis first encountered the elder Hinton's writing in 1878 when he was a schoolmaster in rural Australia. Upon returning to England two years later, Ellis became intimate with James Hinton's surviving family. Ellis began reading and organizing James Hinton's unpublished manuscripts for publication and became further entangled with the Hinton family when James Hinton's sister-in-law Caroline Haddon partly financed his medical studies.[15] During this time, Ellis also became a founding member of the Fellowship of the New Life, a socialist group that co-evolved alongside the Fabian Society. Ellis accompanied Margaret Hinton, James Hinton's widow, and Caroline Haddon to the first meeting of the Fellowship; after the division between the Fellowship and the Fabians, Haddon continued with the Fabians.[16] The Fabians were concerned with reforming economic and social policy, while the Fellowship focused more on the improvement of the individual. A member of both societies, Joseph Oakeshott, described the difference as one of 'Practical versus Idealist Socialism', claiming the two were inextricably linked:

> The work of the Fabian Society is to agitate for and bring about an economic and political reconstruction of society […]. But the specific work of the Fellowship is (1) to lay down what it considers the true principle of human relationships and social organization; (2) to point out, how far the present social system is out of harmony with this idea; (3) to call on those round about it to work for the immediate realization, as far as may be possible, of the accepted ideal in daily life; and thus supply the necessary ethical impulse to the economic and political movement.[17]

It is not difficult to see how Hinton's hyperspace philosophy would appeal to those who were committed to working for the 'immediate realization' of the 'accepted ideal in daily life'. Two years previous to this statement, Hinton had written that 'the way to know' how to live and whom to serve is to 'get somehow a means of telling what your perceptions would be if you knew, and act in accordance with those perceptions'.[18] It is through individual acts of will-to-belief that broad social change is enacted.

While it is not clear if Hinton attended Fellowship meetings himself, it is apparent that his close friends and family moved in London's progressive circles. His father's philosophy was read and debated avidly in these circles, and Hinton's own bigamy scandal was a topic of discussion both in private correspondence and in the meetings of Karl Pearson's Men and Women's Club.[19] Matthew Beaumont identifies 'a politics of fellowship' within late-Victorian progressive circles that he describes as underlying 'the structure of feeling specific to feminist and socialist-feminist utopianism in the late nineteenth century'.[20]

---

15 Ellis, *My Life*, 143–44.
16 Grosskurth, *Havelock Ellis*, 68. Grosskurth notes that Haddon's paper, 'The Two Socialisms', presented a Fabian meeting in March of 1884 marks the first time that the word 'socialism' appears in the records of the Fabian Society. See also N. I. MacKenzie and J. MacKenzie, *The First Fabians*, 1–45.
17 Oakeshott, 'Practical', 13.
18 Hinton, *A New Era*, 74.
19 Nottingham, *The Pursuit*, 50.
20 Beaumont, 'A Little Political World', 222.

Members of the Fellowship, the Men and Women's Club and similar groups dreamed of utopian futures of equal class and gender relations while attempting to practice 'alternative personal relations that, at a local level, were a proleptic form of the universal collective of the future'.[21] The 'alternative personal relations' practised by the Fellowship were broadly divided into non-traditional sexual relationships and acts of public service.

It is not surprising that James Hinton's ideas continued to appeal to Ellis and other participants in the 'politics of fellowship'. In Ellis's interpretation, '[James] Hinton sought to supplement' his 'altruistic concept' with the ideal of '*service*. By sacrifice he had meant the willing acceptance of pain, all thought of self being cast out; by service [...] he meant acceptance of pleasure also'.[22] His 'altruistic' justification of pleasure was popular amongst some idealist socialists who were concerned with establishing equality between the sexes through the demystification of sexuality and birth control. Ellis himself was drawn to James Hinton's allowance for pleasure within an ethics that treated morality as 'a relation which must be fluent, which cannot be rigid', as well as liberating 'that tendency to impulse and the free play of passion'.[23]

As noted in the first chapter, there was a darker side to James Hinton's philosophy. In the small, overlapping network of British progressives that comprised the Fellowship, the Fabians and other groups such as the Men and Women's Club, gossip as well as ideas travelled quickly. Ellis soon began to uncover various hints of scandal surrounding James Hinton and his followers, referred to by contemporaries as 'Hintonians'. In addition to his wife, Margaret, it appears that James Hinton had many mistresses, including his sister-in-law Caroline Haddon and family friend, Mary Everest Boole, widow of mathematician George Boole and Charles Howard Hinton's future mother-in-law.[24] While Boole, Haddon and other Hintonians staunchly defended James Hinton against these accusations, troubling reports of his predatory behaviour began to surface: New Woman novelist and Fabian member Emma Brooke wrote to Ellis in 1885 to warn him that nothing could be 'more injurious to the liberty of woman' than James Hinton's tenets enacted. Writing, of her own experience with James Hinton, Brooke described how she was once thrown into his company for an extended period of time at a mutual friend's house. After fending off his advances for a number of days:

> At last things came to such a pass that I was obliged to tell him I loathed him and I coupled this with some caustic remarks as to the unmanliness of his conduct. He then told me that he was aware I disliked his attentions and he had thereby the hypocrisy to add that he wished

---

21 Ibid., 222. That Hinton's hyperspace philosophy was popular amongst such communities is evidenced by Hiram Barton, who, in response to a 'Mathematical Games' column in *Scientific American*, notes that Francis Sedlak, an early member of the Whiteway colony in Gloucestershire, worked Hinton's cube exercises earnestly. See Gardner, *Mathematical Carnival*, 52–53.
22 Ellis, 'Hinton's Later Thought', 394, original emphasis.
23 Ibid., 395.
24 Adding to the scandalous nature of James Hinton's sexual relationship with Caroline Haddon was the continuing taboo against relationships with siblings-in-law, which considered them incestuous. This taboo is reflected in the legal prohibition against a widower's marriage to his sister-in-law, which was only overturned in the 'Deceased Wife's Sister's Marriage Act' of 1907.

to teach me that duty and loveliness of yielding myself to 'others' needs' and wishes, and of over-coming all 'self-regarding impulses'.[25]

Here we have a disturbing example of James Hinton's philosophy of altruism and free-love in practice. 'The worst of it is', Brooke continued, 'that my experience as regards [James] Hinton by no means stands alone; he was in the habit of persecuting young girls'.[26] Ellis uncovered other examples of James Hinton's pseudo-philosophical rationalization of sexual harassment. For example, he was a proponent of female nudity: 'the nakedness of women is a point he insists on a great deal. He puts it rather one-sidedly; he doesn't say much about men going naked', Ellis explained in correspondence with Olive Schreiner.[27] Schreiner and Ellis frequently discussed James Hinton's ideas and the Hintonians, and she wrote to him after hearing of two separate instances in which James Hinton was witnessed undressing and handling his own daughters, once with Charles Howard Hinton present as well.

In spite of the rumours surrounding the Hinton family, Charles Howard Hinton's work was growing in popularity in the 1880s. As well as his textbook, *Science Note-book* of 1884, the texts of his first series of *Scientific Romances* were appearing in pamphlet and volume form; these publications received favourable reviews in *Mind* and *Nature*, among other periodicals.[28] In addition to the growing interest in four-dimensional geometry and physics in these scientific periodicals, the ethical possibilities of accessing the fourth dimension – as expressed in Hinton's hyperspace philosophy – were being discussed amongst members of the Fellowship of the New Life, the Men and Women's Club and the Fabian Society.[29] Once the scandal of Hinton's bigamy trial broke, this early success was cut short in Britain and, unable to find work, Hinton left for a teaching post in Yokohama, Japan. Both Ellis and Schreiner were intimately involved in the details of Hinton's trial and subsequent departure, and Schreiner attended the trial.[30]

Although his name became anathema in Britain, Hinton's ideas continued to resonate across progressive circles. For an example of how Hinton's hyperspace philosophy – with its championing of the broadly equalizing potential of higher space – appealed to the radical left, we can consider Edward Carpenter's published account of his travels across Sri Lanka and India, *From Adam's Peak to Elephanta* (1892). This text was controversial

---

25 Letter from Brooke, 5 August 1885, in the Papers of Havelock Ellis, 38–39. It appears that Brooke was originally put into contact with Ellis through Olive Schreiner; see '*My Other Self*', ed. Dranzin, 53–56. Brooke wrote to Pearson about this incident as well. See Porter, *Karl Pearson*, 136.
26 Letter from Brooke, 5 August 1885, in the Papers of Havelock Ellis, 40.
27 H. Ellis to Schreiner, Olive Schreiner archive, Harry Ransom Research Center, University of Texas. Quoted in Brandon, *The New Women*, 61.
28 No copies of *Science Note-book* appear to be extant.
29 Porter notes, for example, that 'Olive Schreiner corresponded with Maria Sharpe about the fascination of four-dimensional space' (193).
30 See '*My Other Self*', and *Olive Schreiner Letters*, ed. Rive, 106–14. Grosskurth notes that H. Ellis acted in some capacity as Maude Weldon's physician; she tentatively speculates that he may have delivered Weldon's twins (102).

because of its expression of support for Indian nationalists, but what is of particular interest here is how Carpenter engaged with what he calls the 'cosmic consciousness'. Carpenter attempted to describe, 'as far as I can, in my own words, and in modern thought-forms', the teachings of the 'Indian Gurus' encountered during his travels:

> Yet there may be an inner vision which again transcends sight, even as far as sight transcends touch. It is more than probable that in the hidden births of time there lurks a consciousness which is not the consciousness of sensation and which is not the consciousness of self – or at least which includes and entirely surpasses these – a consciousness in which the contrast between the *ego* and the external world, and the distinction between subject and object, fall away.[31]

This state of higher, or 'cosmic', consciousness is one in which the boundary between self and other is dissolved, what some psychologists before Freud termed the 'secondary consciousness'. Carpenter argued that 'the idea of another consciousness, in some respects of wider range than the ordinary one, and having methods of perception of its own, had been gradually infiltrating itself into Western minds'. Along the same lines, Carpenter continued, 'science has been familiarising us with, and [...] bringing us towards the same conception – that, namely, of the fourth dimension'.[32]

The analogy Carpenter deployed in his attempt to represent the social and ethical implications of realizing this cosmic consciousness reveal a familiarity with the discourse of hyperspace philosophy:

> As a solid is related to its own surfaces, so, it would appear, is the cosmic consciousness related to the ordinary consciousness. The phases of the personal consciousness are but different facets of the other consciousness; and experiences which seem remote from each other in the individual are perhaps all equally near in the universal. Space itself, as we know it, may be practically annihilated in the consciousness of a larger space of which it is but the superficies; and a person living in London may not unlikely find that he has a backdoor opening quite simply and unceremoniously out in Bombay.

This space-annihilating cosmic consciousness is interchangeable with the four-dimensional perception, as are the implications of its realization. Individual consciousness is subsumed within the universal, and from the perspective of the cosmic consciousness, Britain and India are one. 'This sense of Equality, of Freedom from regulations and confinements, of Inclusiveness, [...] belongs of course more to the cosmic or universal part of man', he concludes.[33] It is this 'universal part of man' that both Carpenter and Hinton viewed as the foundation for social equality. It is not coincidental then, that the socialist revolutionaries who invade the British political scene in Ford Madox Ford and Joseph Conrad's *The Inheritors* (1901) are from the fourth dimension.

---

31 Carpenter, *From Adam's Peak*, 155, original emphasis.
32 Ibid., 160–61.
33 Ibid., 162.

Gendered Temporality

The temporal order of capitalism is also disrupted here: Carpenter's argument is privileging a 'primitive' sense of cyclical time and space in which all the phases of the individual's life are subsumed within an ever-present consciousness. Thus, the preference for 'a privileged linear model [over] an unprivileged cyclical model of history' is overturned.[34] Patricia Murphy has explored how Victorian perceptions of this binary were often gendered: in much of the fiction of the nineteenth century, 'a male character's association with linear time becomes a marker of progress, civilization, and modernity, whereas a female character's connection to cyclical time represents stasis, chaos, and anachronism'.[35] This linking of the female with the cyclical was pathologized during this period as well; with her monthly menstrual cycle, the female is somehow connected to nature in a way that the male is not. Women and the labouring poor are bound to repetition and the chaos of nature, professional men to progress and culture. The churning network of assumptions that underpins much of Victorian social thought is made explicit in Havelock Ellis's proto-feminist but essentialist argument that

> while women have been largely absorbed in that sphere of sexuality which is Nature's, men have roamed the earth, sharpening their aptitudes in perpetual conflict with Nature. It has thus come about that the subjugation of Nature by Man has often practically involved the subjugation, physical and mental, of women by men.[36]

For Ellis, these essential differences between the sexes support what he argues is female superiority in ethics and political governance. Females, according to this line of reasoning, are more closely aligned with nature and therefore exhibit 'organic conservatism'. In applying these 'zoological facts' to politics, Ellis found that organic conservatism may

> often involve political revolution. [...] Socialism and nihilism are not, I believe, usually regarded by politicians as conservative movements, but from the organic point of view of the race they may be truly conservative, and as is well known, these movements have powerfully appealed to women. [...] Taking a broad view of the matter, it seems difficult to avoid the conclusion that it is safer to trust to the conservatism of nature than to the conservatism of Man.[37]

Here Ellis was on the side of 'organic conservatism', which – rather than being politically conservative – is in fact politically revolutionary.[38]

While late twentieth- and twenty-first century critics tend to identify Ellis and James Hinton as anti-woman (if they consider them at all), both men (and many of their

---

34 Murphy, *Time Is of the Essence*, 24.
35 Ibid., 24.
36 Ellis, *Man and Woman*, 395.
37 Ibid., 370 and 397.
38 This brand of proto-feminism was alive and well into the twentieth century; in fact, it underpins the psychologist William Moulton Marston's rationale for creating the comic book character, Wonder Woman. See Lapore, *The Secret History*.

contemporaries) thought of themselves as what we would now call feminists. George Peard, a personal acquaintance of James Hinton, recalled that 'we never attended a meeting on any Woman's Question, educational or other, without meeting James Hinton'.[39] He was particularly concerned with ending prostitution, and it is clear that James Hinton saw himself as a progressive when it came to any 'Woman's Question': he is said to have declared himself the 'saviour of women'.[40] When first introduced to James Hinton's philosophy by Ellis, Olive Schreiner wrote enthusiastically that she no longer despaired that 'woman will have to save woman alone'.[41] It was only when gossip mounted and the younger Hinton's bigamy scandal broke that Schreiner vehemently declared herself anti-Hintonian.

Charles Howard Hinton did not, himself, explicitly address any questions concerning women. However, he was writing out of a social milieu in which feminist issues were debated continually, along with socialism and science. It is part of my aim in the present chapter to demonstrate how Hinton's challenge to a linear, irreversible temporal model can be understood as having radical implications for the gendered subject. As Elizabeth Grosz observes, 'representations of space and time are in some sense correlated with representations of the subject. [...] If space is the exteriority of the subject and time its interiority, then the ways this exteriority and interiority are theorized will effect [sic] notions of space and time'.[42] In both *Stella* and *An Unfinished Communication*, Hinton explored the connection between representations of the subject and time and space. In the second series of *Scientific Romances* we can observe a marked turn inward, toward an attempt to imagine the four-dimensional self.

I have begun this chapter by re-examining the context of Hinton's work because it is necessary to fully appreciate the ways that his hyperspace philosophy influenced and was influenced by his contemporaries. Though he is now an obscure figure, in his personal life Hinton crossed paths with a number of writers, scientists and social activists. These people were reading his hyperspace philosophy and having conversations about the fourth dimension with Hinton and amongst themselves. To look into Hinton's past and examine his work is to uncover another facet of the shared literary and cultural movement we call modernism. Hinton, like his contemporaries, was writing to and reacting against the previous generation which, for him particularly, included his father. His theory of the fourth dimension can be read as part of this reaction, this attempt to transcend the past. Hinton most explicitly grappled with the problems of modernity through his experimental fictions of the second series, *Stella* and *An Unfinished Communication*.

Stella as Experiment

The timing of the composition of *Stella* is important. Originally published in 1895, this text appears to have been composed after Hinton's bigamy conviction. Certain elements

---

39 Peard, 'The Hintons', 269.
40 E. Ellis, *Three Modern*, 17.
41 Schreiner to H. Ellis, 2 May 1884, in Dranzin, '*My Other Self*', 44.
42 Grosz, *Space, Time, and Perversion*, 99.

of the story parallel Hinton's life: Stella's husband accepts an appointment in the Far East and they both travel around the world. It is implied that their emigration from England is forced because of the couple's limited funds and Stella's strange optical status: she is an invisible woman. Thus, like Hinton and his (legal wife) Mary, Stella and her husband Hugh Churton leave England for foreign appointments under duress.

Predating H. G. Wells's *The Invisible Man* by over a year, *Stella* is the first story of human invisibility induced by scientific (as opposed to magical) means. Stella was made invisible by an older man – her legal guardian, Michael Graham, who died when she was 17 years old. Graham found the means of inducing Stella's invisibility through his research into 'the border land between chemistry and physics'.[43] However, his reasons for causing Stella's invisibility, rather than the pseudo-scientific explanation of it, are of primary concern in this chapter. In addition to being a scientific investigator, Graham is described as a philosopher whose ideas are similar to James Hinton's. Graham made Stella invisible as an experiment to see how a human being would react when the opportunity for self-regard was removed.

Stella served as an experimental fiction in which Hinton was able to test out his own ideas and critique his father's philosophy. After obliquely referring to his bigamous marriage to Maude Weldon in *A New Era of Thought*, Hinton concluded that 'still, it does seem that, as an ideal, the absolute absence of self-regard is to be aimed at. [...] And if this law of altruism is the true one, let us try it where failure will not mean the ruin of human beings'.[44] If, as it seems to be implied here, Hinton's bigamous marriage was a ruinous experiment in his own life, *Stella* is an attempt to explore 'the absolute absence of self-regard' in his fiction.

Graham's decision to experiment on a female subject as opposed to a male one is informed by many of the Victorian assumptions about gender I have touched upon. These assumptions are, for the most part, carried through without question in *Stella*; however, one result of this experiment in fiction is the acknowledgement of the double-bind placed upon Victorian women.[45] The story itself – partly as a result of the structure of the narrative – is ambivalent. The structure of *Stella* adheres to a gothic pattern: there is an unnamed external narrator who frames Churton's report of his uncanny encounter with an invisible woman whom he initially mistakes for a ghost. In the opening pages of the novella, the unnamed narrator describes how he, Hugh Churton, and a mutual friend, Frank Cornish, met when they were students. The narrator explains their circle as one of young men primarily interested in 'a very worthless and corrupt side of London life'.[46] The nature of this corruption is alluded to by their love of Zola, whom they read avidly. All three young men are members of 'a band of pupils' who surround a Victorian sage, Dr Forysth, who appears to be partly modelled on Ruskin. Cornish is a neurotic and brilliant scientist whose alcoholism brings him near death at least twice. Churton is the exact opposite of Cornish, belonging to the class of men

---

43 Hinton, *Stella, Scientific Romances*, 17.
44 Hinton, *A New Era*, 91.
45 See also Throesch, 'The Difference between Science and Imagination?'
46 Hinton, *Stella, Scientific Romances*, 2.

characterized like many Englishmen by a disinclination for a life of study amounting to incapacity. They were the kind of men who have the habit of being elected captains of their football or cricket teams when young, and of being people to be considered afterwards, but of a mental disposition which makes it an episode of decidedly healthy tendency for them to attempt to be selected for administrative posts by competitive examination.[47]

The unnamed narrator is critical of Churton in both the opening and closing frames of the novella. His description of Churton above echoes criticism of the codification of athletic activities in Oxford during the 1870s; Ruskin's Hinksey road project was in part a reaction to this 'new cult of sports'. In his biography of Ruskin, Tim Hilton describes the anti-athletic element of the Aesthetic movement, claiming that Oscar Wilde 'led the first generation of Oxford undergraduates who mocked sporting men while proclaiming themselves the initiates of art'.[48]

The narrator's critique of the sporting type would align him with those who opposed the introduction of organized sports into Oxford culture. He seems to possess a more sensitive, aesthetic nature than Cornish or Churton, and he is a writer. In the end, it is the narrator who reads Graham's manuscripts and, having 'mastered their contents' (as opposed to Churton with his 'more than average English incapacity for ideas'), is able to share Stella's story with the world. He writes a book about her, unbeknownst to Churton who, in Stella's words, 'doesn't read anything except the newspapers' and wouldn't 'mind anything written in a book'.[49]

Significantly, the narrator links Churton's bullying, anti-intellectual temperament with British imperialism. The administrative post for which Churton is competing is in the Civil Service, and the narrator harshly critiques British colonial rule and the 'type' of colonial administrative official that Churton represents:

> The natives of India look on our ascendancy with a dumb resignation to the designs of an inscrutable providence, which so often lets rude force hold sway over all the gentler virtues. In the mining company with which I became connected […] there are numbers of native employés [sic], excellent men, most admirable in every private relationship; but they all occupy subordinate positions. We have to put over them some low-lived swearing Englishmen, with one-tenth of their mental ability, if we want the work done. There is something the Hindoos [sic] lack and which Churton possessed in abundance.[50]

Churton thus becomes a representative of masculine Victorian temporality; an ambassador for capitalism and empire, he is the authoritarian figure who imposes linear, 'progressive time' onto the feminized colonial subject.

While the unnamed narrator frames the story, the bulk of the narrative is in Churton's voice. However, the framing critique of Churton casts doubt on his subsequent narrative, his behaviour toward Stella and his interpretation of Graham's

---

47 Ibid., 8.
48 Hilton, *John Ruskin*, 293.
49 Hinton, *Stella, Scientific Romances*, 107.
50 Ibid., 9.

philosophy. Churton's narrative begins with his arrival at the recently deceased Michael Graham's house to settle his estate as a favour to Cornish, who is abroad. Graham, Frank Cornish's uncle, was a wealthy bachelor, who lived in seclusion in his manor house on the Yorkshire moors. Unbeknownst to his relatives, he had become the guardian of the orphaned Stella, and, after his death, Stella is left alone in the house with two servants. It is Churton who first discovers Stella's presence in the house, and after realizing that she is a living, invisible girl rather than a ghost, he sends for Cornish's mother to act as a chaperone. Before Mrs Cornish is able to arrive, however, a fraudulent spiritualist kidnaps Stella and tricks her into performing at his séances. Churton tracks down Stella and rescues her, marries her and takes her to China. While in China, Churton becomes increasingly dissatisfied with Stella's invisibility, and upon learning that her strange optical status is maintained only by regular consumption of a chemical compound, he convinces her to become visible again. After Churton's triumph, they return to England, where Stella gives birth to a son.

During Churton's time at Graham's estate, it is his job to read and organize Graham's philosophical manuscripts. Churton misreads Graham's ideas, which are explicable only to someone who has an understanding of the fourth dimension. However, Hinton never explicitly mentioned the fourth dimension in this novella; Churton's failure to acknowledge it is only implied. Churton relays Graham's writing:

> Conceive now a solid with a great number of faces. These faces are separated from each other by bounding lines, which are the edges of the solid. Suppose these faces to have the power of reasoning and reflection. Let them come to a consciousness of each other. They know first of all of their relations – to explain these they conceive of themselves and those about them to have substance. This substance, however, is what we call plane surface. Now we know that what they call substance is really a relation. Not knowing of the solid, they think that superficial substance is the ultimate substance. But this substance of theirs is really in the same relation to the solid as their lines are to their substance.[51]

He then offers his own analysis:

> This is an instance of Michael Graham's *'path.'* The objection to it of course is, that we come to a dead wall. There isn't anything beyond a solid substance of which the solid is a relation in the sense used above.[52]

However, for believers in the fourth dimension, there *is* something beyond a three-dimensional solid. Churton fails to recognize the dimensional analogy implied here, and, as a result disregards Graham's ideas as part of 'a bankrupt philosophy'.[53]

Stella is therefore a multivocal text. Graham's writings are relayed verbatim by Churton, who offers his own interpretation and critique. Stella also explains *her* understanding (which

---

51 Ibid., 53.
52 Ibid., 53, original emphasis.
53 Ibid., 16.

appears to be aligned with Hinton) of Graham's philosophy, which is in turn relayed to the reader by Churton and the framing narrator. The narrator offers a meta-commentary on all events and characters. These multiple viewpoints are often in conflict with each other, making it impossible to tease out where exactly Hinton's sympathies lie. The expectation – based on a reading of the first series of *Scientific Romances* – that *Stella* will be a straightforward allegory of hyperspatial awareness, is therefore subverted by the competing voices within this text. The second series is more 'literary' in this sense. In my reading of *Stella*, I do not seek to 'decode' this text; rather, I will analyse these competing narratives.

Untimely and Invisible

That Hinton intended both *Stella* and *An Unfinished Communication* as experiments on the 'untimely', four-dimensional individual is made explicit in a letter to William James in 1895:

> In working with the 4 dimensional space hypothesis it became evident that the method of forming the working intuition of this extended space was to use time as the fourth dimension. To assume that matter had another dimension which is experienced by us as duration. [...] The difference in the view of the world, the aspect which things come to wear from this assumption incorporated and made familiar[,] I have expressed in 'Stella,' partially, & more fully in 'An Unfinished Communication'.[54]

The individual with a four-dimensional perspective is able to see that time is in fact an illusion; it is simply the way the limited three-dimensional consciousness encounters hyperspace.

Graham rendered Stella invisible so that she might be better equipped to experience eternity. This is not the eternity described by the Christian doctrine of the immortality of the soul, or 'time carried on and on', as Stella describes it, where 'our souls leave our bodies and are in the presence of God'. Graham's understanding of eternity is explained by Stella in the same vocabulary and imagery that Hinton and Carpenter used when discussing 'higher consciousness': 'find[ing] your eternal self is not to find yourself apart and separate, but more closely bound to others than you think you are now. You learn yourself in finding yourself linked with others'.[55] While Carpenter only implicitly linked his cosmic consciousness with time, Hinton made it clear that time must be subsumed into four-dimensional perception. To find one's eternal self is to annihilate time and space, just as in 'What Is the Fourth Dimension?' Hinton implied that the sensation of duration is merely the way in which the limited, three-dimensional consciousness can encounter the fourth dimension of space.

---

54 Hinton to W. James, 19 October 1895, 3. James Family Papers, in Houghton Library, Harvard University, bMS Am1092 (368–77).
55 Hinton, *Stella*, *Scientific Romances*, 29.

This eternal self is *un*timely because time does not exist 'when', as Stella explains, 'you learn yourself truly':

> If you feel eternity you will know that you are never separated from any one with whom you have ever been. You come to a different part of yourself each day, and think the part that is separated in time is gone. But in eternity it is always there.[56]

This eternal self – which is the same as Carpenter's cosmic consciousness – subsumes individual selves. Stella's guardian, Graham, decided that it is through extracting the social aspects of selfhood that the eternal self is found. 'We must remember that the [individual, non-eternal] self is a relative term', Churton reads in Graham's papers; 'the higher self is that through which these conflicting selves exists, through which each has its individuality'.[57] This higher self is like the Persian king, through whom enlightened valley-dwellers such as the student can become 'conscious of being true selves, the oneness of all of them lying in the king, but each spontaneous in himself and absolute will, not to be merged in any other'.[58] Graham attempted to access this higher self through his experiment on Stella. Unfortunately for Graham and his experiment, Stella – as a Victorian woman – is expected to become 'merged' in another, namely her husband.

Rather paradoxically, Graham planned to allow Stella to develop into a 'true' self by totally annihilating her sense of self. By rendering Stella invisible, he was attempting to prevent her from developing regard for herself as an individual; this lack of self-regard, he believed, would better enable her to access the higher self. Here is where Stella's gender becomes important, as Graham explained:

> The body and the moral sense are intimately connected. A Habit is morality of some kind become bodily. [...] In its earliest movements a child incorporates, in its very constitution, tendencies which we afterwards, alas! and rightly then, recognise as part of its nature, which it may learn to restrain, from which it can never escape. Think of a little girl, almost from the time when she can first see, creeping up to a glass and looking at herself, decking herself with a ribbon or a string of beads. We allow this thinking about self, this vanity, to become incorporated in the female child; all subsequent education simply leads it to disguise itself, we can never eradicate it.[59]

Graham argued that the supposed feminine tendency to develop self-regard through objectification is in fact learned. Similarly, nonessentialist writers such as Schreiner and Wilde were arguing that 'the only thing that one really knows about human nature is that it changes'.[60] If social conditions were to change, Wilde wrote in 'The Soul of Man Under Socialism', then human nature would follow. The rationale for Graham's experiment was this belief in the social construction of human 'nature'; his approach of experimenting on an individual girl was also aligned with the 'fellowship politics' of the Fellowship of

---

56 Ibid., 30.
57 Ibid., 107.
58 Hinton, 'The Persian King', *Scientific Romances*, 128.
59 Hinton, *Stella, Scientific Romances*, 15–16.
60 Wilde, *Complete Works*, 1100.

the New Life, which subordinated large-scale social activism to personal renewal and development.⁶¹ In opposition to Graham, Churton is aligned with an essentialist viewpoint: upon reading the passage quoted above, he 'closed the book, pitying the poor child who had been employed posting up the ledger of a bankrupt philosophy, when she might have been playing tennis or dancing'.⁶² In Churton's later misreading of Graham's rationale for experimenting on a female subject, the difference between social constructionism and biological essentialism is subtle but important. Churton explains how

> Michael Graham had resolved to try practically what direction the activities of the soul took when the self-regarding impulses were denied the opportunity of existence. A boy cares about eating and drinking and getting things. You could not deprive him of these self-centred activities of his, without [killing him] [...]. Instead of a boy he experimented on a girl, for a girl's self-love is concerned with being looked at – it is in producing an effect on others that her self-love is gratified. By taking away visible corporeality from Stella, he took away the means of living for herself.⁶³

In Churton's reading, a woman's only 'means of living for herself' is through objectification. He does not register Graham's observation that feminine self-validation through vanity is a *learned* activity, and not an *essential* part of female nature. Rather, Graham's intention was to remove Stella from an economy where the value of women is determined by their attractiveness as objects. His hope seemed to be that, deprived of her only capital, Stella would become a model socialist.

Here is where Stella's personal development takes on political implications. Graham's view of the possible configurations for society was binary: the first possibility which, he argued, is how his society was currently organized, is based on 'self-interest' on the part of each individual. The second, which he preferred, is an inversion of the individualist model:

> [Graham] asserted that the forms of life, on the hypotheses of acting for others and acting for self, are almost the same. The transition from the operation of one principle to that of the other would imply no violent outward change, but a difference in the working of each part.⁶⁴

This was the methodology of the Fellowship of New Life: rather than seek change through legal reform like the Fabians, the Fellowship focused on changing 'the working of each part' of society, or each individual subject. Graham's socialist model of acting for others necessitates a broader view that transcends the individual's placement in time and space. Graham attempted to achieve this transcendence through invisibility, which he thought would allow Stella to actually *feel* in accordance with his philosophy in the same way that Hinton believed his readers could learn to feel the spatial relations of

---

61 See Summers, 'The Correspondents', 167–83.
62 Hinton, *Stella*, 17.
63 Ibid., 48–49, emphasis added. Note the exact reiteration of the phrase 'self-regarding impulses' that Brooke places within quotation marks in her letter to H. Ellis.
64 Ibid., 47.

each individual cube through 'casting out the self'. For both Hinton and Graham, it was necessary to actualize this feeling because 'we each of us have a feeling that we ourselves have a right to exist. We demand our own perpetuation. No man, I believe, is capable of sacrificing his life to any abstract idea'.[65]

In order to overcome the self-regarding impulses, each individual must be able to intuit higher space so they can feel 'all mankind as physical parts of one whole', and understand that 'our apparent isolation as bodies from each other' is in fact a fallacy.[66] Though he never explicitly used the word 'socialism', the implications of Hinton's hyperspace philosophy and Graham's experimentation are clear. By radically reorganizing human experience of time and space, one individual at a time, the entire social body will undergo a transition away from individualist capitalism and the model of linear, irreversible time that underpins it:

> To get a knowledge of humanity, we must feel with many individuals. Each individual is an axis as it were, and we must regard human beings from many different axes. And as, in learning the block of cubes [...] is the means by which we impress on the feeling of different views of the block; so, with regard to humanity, it is by acting with regard to the view of each individual that a knowledge is obtained. That is to say, that, besides sympathizing with each individual, we must act with regard to his view; and acting so, we shall feel his view, and thus get to know humanity from more than one axis. Thus there springs up a feeling of humanity, and of more.[67]

Unfortunately for Stella, like Ruskin's paradoxical invisible man, she is stuck within her own double bind. Defined by two men in opposition with each other, Stella has to choose between being Graham's socialist emblem or Churton's wife. Churton positions Stella's return to visibility as a reclamation of her self as an autonomous being; however, in Hinton's logic, this is the point where Stella loses her sense of higher self and rejoins the capitalist order in which women are subordinated as reproducers. After they are married, Stella eventually gives in to Churton's pressure and becomes visible again, even though, as Churton himself acknowledges, Stella 'really felt as if being seen was – she felt about it as a well-bred lady would about exposing more of her person than society permits'.[68] The irony of Stella's loss of self through reclamation of her visibility is demonstrated when 'she received quite an ovation from the ladies of Hong-Kong [who] admired the complete way in which Stella had put down my monstrous disposition to jealousy'.[69] While she was invisible, Stella wore a veil in public in order to disguise her strange (lack of) appearance. The women who applaud Stella misread her unveiling as her successful self-assertion over her husband's jealousy rather than her actual total submission to his wishes. After Stella becomes visible, Churton learns of her substantial inheritance and they are able return to England where she gives birth to a son; thus, the return to order and the continuance of Churton's name and property is assured.

---

65 Hinton, *A New Era*, 71.
66 Ibid., 67.
67 Ibid., 77.
68 Hinton, *Stella, Scientific Romances*, 86.
69 Ibid., 105.

Hinton did not end the novella here, however. The conclusion is the narrator's description of his conversation with Stella, after he has heard Churton's version of the story. The narrator's sympathies appear to lie with Stella, and Churton is not present at the conversation that provides the main content of the conclusion. The conversation casts a shadow of ambiguity over the traditional happy ending of Churton's narrative. When the narrator observes to Stella that 'it has ended happily!', her reply, though affirmative, is hesitant: 'Yes, Hugh [Churton] put everything to rights; but I feel as if I had forgotten something, as if we had all forgotten. [...] I cannot quite be happy often'. Stella realizes that something is missing from her life, but she is unable to articulate anything but its absence. Her scope for action is thus limited by her inability to define her feeling of loss and what exactly she has lost. When the narrator asks her, 'But what can you do?', she can only respond: 'That is the sadness. I don't know how to do what [Graham] wanted'.[70] Stella positions her problem as a loss of the ability to fulfil Graham's wishes. The cause of her dissatisfaction, which she is unable to identify, appears to stem from the opposition of Churton's desires with Graham's, of capitalist and socialist values.

Nietzsche, the 'Disadvantages' of the Past and *An Unfinished Communication*

Given the failure of the experiment with Stella, it is not surprising that we see Hinton turn even further inward with his final text in the second series, *An Unfinished Communication*. This text is where I will turn for the remainder of this chapter. Of particular interest in *An Unfinished Communication* is Hinton's treatment of time; the narrative ends with a four-dimensional vision that is markedly similar to Nietzsche's idea of eternal recurrence. Like Nietzsche, Hinton used his vision of eternal recurrence to liberate the individual subject from 'the temporal logic of modernity', where 'time and history are [viewed as] problems to be overcome'.[71] It may seem strange to shift from a discussion of fin-de-siècle socialists and literati to Nietzsche; however, numerous Victorian progressives and socialists appreciated Nietzsche. Havelock Ellis's wife Edith even went as far to claim that James Hinton was an intellectual precursor to Nietzsche.[72] Ellis himself was one of the earliest English commentators to review Nietzsche. Nineteenth-century British socialists had little problem reconciling Nietzsche's emphasis on the individual will with their collectivist ideals:

> The introduction of collectivism in some areas was not seen as restricting the full exercise of individual freedom in others. Indeed one of the essential claims of socialism [...] is its unique ability to reconcile the two ideas; the future socialist form of society being the only one in which the rival claims of individual self fulfilment and collective solidarity can be reconciled.[73]

---

70 Ibid., 106–7.
71 Rampley, *Nietzsche, Aesthetics and Modernity*, 148.
72 See E. Ellis, *Three Modern Seers*.
73 Nottingham, *The Pursuit of Serenity*, 239.

That Nietzsche held appeal 'for those who rejected, or felt rejected by, the liberal commercial society' is confirmed by Max Nordau, who linked Nietzsche along with other 'ego-maniacs' and 'degenerates' such as Wilde and Ibsen.[74] Nietzsche's transvaluation of values, his model of eternal recurrence and his ideal of the *Übermensch* resonates within Hinton's final romance; it is here that Hinton was able to dramatize an individual's encounter with the fourth dimension as a discovery of the will-to-create.

The key to the narrator's four-dimensional experience in *An Unfinished Communication* is his relation to time. As I mentioned at the beginning of this chapter, Hinton's hyperspace philosophy reduces time to an effect of four-dimensional space. For Hinton, time was merely a useful tool; it is simply the way in which the untrained human mind encounters the fourth dimension of space. In a letter to William James, Hinton expanded on this theory:

> We must get sensations of the higher. As you know I think that it is to be got by fusing extension and duration together making that unity which we may call the process thing – a higher material existence, simultaneous actuality, which we apprehend as consecutive, given in fugitive nows.[75]

In Hinton's hyperspace philosophy, each 'fugitive now' is to the human mind what the two-dimensional 'slice' of a solid would be to a flatlander. Hinton's interest in transcending the sensation of duration as discontinuous ('given in fugitive nows') is relevant later in my discussion of William James's stream of consciousness. In the present chapter, what is of interest is how Hinton attempted to express his idea of 'simultaneous actuality' as a means of overcoming the tyranny of linear, 'objective' time in *An Unfinished Communication*. It is this experiment with temporality and eternal recurrence that marks his final scientific romance as distinctly modern.

Unlike *Stella*, with its multi-vocal, gothic structure, *An Unfinished Communication* appears at first glance to be a more straightforward, univocal text. It is narrated by the protagonist, an unnamed man who is wandering through the northeastern United States, trying to escape his past. However, in the eternal recurrence scenes in the closing pages of the book, Hinton drew on his own four-dimensional theories, Nietzschean philosophy and recent innovations in moving-picture technology to disrupt linear narrative time.

Broadly speaking, Nietzsche was also concerned with time and its role in our construction of values, subjectivity and 'truth'. In his essay 'On the Uses and Disadvantages of History for Life' (1873), Nietzsche addressed the necessity of forgetting the past in order to maintain the general health and happiness of humanity: 'He who cannot sink down on the threshold of the moment and forget all the past [...] will never know what happiness is – worse, he will never do anything to make others happy'.[76] According to Nietzsche, modern culture was in the process of being poisoned by an excess of history, of defining itself in terms of the past and judging itself on past values. In a later section

---

74 See Nordau, *Degeneration*, particularly 241–65.
75 Letter from Hinton to W. James, 10 July 1897, 3.
76 Nietzsche, *Untimely Meditations*, 62.

of the same essay, he described the deadening weight of the past on the present: 'In the end, modern man drags around with him a huge quantity of the indigestible stones of knowledge'.[77] One way to escape this burden, Nietzsche proposed, was to live unhistorically, experiencing each moment as individual and unconnected to every other moment; this is a process of constant forgetting, which is how, according to Nietzsche, animals experience time.

At the extreme opposite is what Nietzsche called the suprahistorical sensibility, which is the ability to view the entire history of the world simultaneously: past, present and future. Nietzsche acknowledged that both modes are problematic because from the suprahistorical 'vantage point [one] could no longer feel any temptation to go on living or to take part in history, [one] would have recognized the essential condition of all happenings', while the unhistorical view precludes the possibility for progression or the accumulation of culture.[78] Nietzsche was trying to construct a temporal model to counter the will's 'fixation on the past which enslaves it to a moral world-order of guilt, punishment, and revenge'.[79] Here he began to define what he saw as the 'temporal logic of modernity', the will's desire to exact revenge on the past and, by implication, life itself.

The child, in a state of forgetfulness, starts life unburdened and happy, according to Nietzsche, but 'then it will learn to understand the phrase "it was": that password which gives conflict, suffering and satiety access to man so as to remind him what his existence fundamentally is – an imperfect tense that can never become a perfect one'.[80] This early view anticipates a speech Nietzsche's Zarathustra makes about the linear, irreversible model of time:

> Willing liberateth: but how call ye that which putteth even the liberator in chains?
>
> 'Thus it was'; so it is named, the Will's teeth-gnashing and loneliest wailing.
>
> Impotent against that which is done, it is an evil onlooker on all that is past. [...]
>
> It is wroth that time runneth not backwards. 'That which was' is named the stone which it cannot roll away.
>
> Therefore it heaveth stones in wrath and indignation. [...]
>
> Thus, Will, the liberator, became a torturer: on all that can suffer it taketh vengeance because it cannot enter the past. [...]
>
> And now cloud upon cloud rolled over the mind, until at length madness preached:
>
> All things perish, therefore, all things are worthy to perish![81]

For the will that seeks revenge on the past, on that which is also a precondition for life, the only way to overcome the past is through self-annihilation. This sort of attempt to overcome the linear, irreversible model of time is what leads to the decadent nihilism

---

77 Ibid., 78.
78 Ibid., 65.
79 Ansell-Pearson, 'Who Is the *Übermensch?*', 317.
80 Nietzsche, *Untimely Meditations*, 61.
81 Nietzsche, *Thus Spake Zarathustra*, 127.

Nietzsche claimed was plaguing modern Western culture. Nietzsche's concern was with finding a way to teach the will to love the past acts it has willed, *because it has willed them*. It is self-affirmation to acknowledge and celebrate one's creativity as a subject. Nietzsche's eternal recurrence served as a counter to the model of linear, irreversible time. His model of circularity and repetition was intended as a thought-experiment in order to see, as Matthew Rampley explains, 'how "incorporation" of the idea of Eternal Recurrence would *change* and *alter* human thinking and practices'.[82] The intended result of undertaking this thought-experiment was to transcend the limitations of the human to become the Übermensch.

Nietzsche's Übermensch continues to be a problematic term and idea. Its chequered history from evolutionistic superman to fascist ideal is perhaps in part the result of Nietzsche's refusal to pin it down to any consistent and logical definition. Keith Ansell-Pearson argues that 'the Übermensch also refers to the question of "we" in Nietzsche, how author and reader are constituted and transformed in the act of reading', adding that Nietzsche's 'writings are thus to be understood as a gift offered to a humanity which, in the act of reading them with the aid of an "art of interpretation," will constitute itself as an over-humanity'.[83] In this sense, Hinton's writings and the four-dimensional, 'higher' consciousness he tried to cultivate in his readers were further attempts to engender the Übermensch. Both Hinton and Nietzsche wrote toward the future in that their texts function as catalysts for the reader; it is through the act of interpreting the texts that the reader creates the aesthetic sensibility both writers attempted to describe. Formally speaking, their approaches are similar: like Hinton in his *Scientific Romances*, 'Nietzsche offers his readers a dazzling array of styles [...] which offer not one "Truth" but many experiences and many truths'.[84]

While there are as many potential four-dimensional selves as there are *Übermenschen*, only the Übermensch, like the four-dimensional consciousness, is able to withstand the challenge of imaginatively sustaining a temporal order that contradicts the linear, irreversible model of time. Nietzsche's suprahistorical viewpoint is the same perspective available from the fourth dimension: time here does not 'pass'. Rather, one has access to a vision of 'some stupendous whole, where all that has ever come into being or will come co-exists'.[85] At the end of *An Unfinished Communication* the narrator experiences this four-dimensional, suprahistorical vision which, in turn, allows him to observe the eternally recurrent nature of all organic life. For both Nietzsche and Hinton, the idea of eternal recurrence is the spur to a revaluation of values.

Like Nietzsche, Hinton wanted to change the ways in which people think and act. His hyperspace philosophy was in part about establishing an ethics that transcended the notions of good and evil. Underlying this lawbreaking desire is the recognition of the constructed nature of 'truth'. 'What, then, is truth?', Nietzsche wrote: 'A mobile army of metaphors, metonymies, anthropomorphisms, in short [...] truths are illusions of which

---

82 Rampley, *Nietzsche, Aesthetics and Modernity*, 149, original emphasis.
83 Ansell-Pearson, 'Who Is the *Übermensch*?', 323 and 324–25.
84 Ibid., 314.
85 Hinton, 'What Is the Fourth Dimension?', *Scientific Romances*, 24.

we have forgotten that they are illusions'.[86] Likewise, Hinton's narrator recognizes that words and the concepts attached to them are a 'series of tricks and devices whereby they [university professors] teach a man knowing nothing of reality to talk of it as if he did'.[87]

What is interesting about the narrator's development in *An Unfinished Communication* is that he begins by lamenting his inability to access objective reality, the Kantian thing-in-itself, but through his experience of four-dimensional consciousness at the end of the text he learns to celebrate the world-creating activity of his own mind as liberating rather than limiting. Like the celebration of the Persian king's will-to-passivity in the first series, Hinton ends the second series with a celebration of the will-to-create, or *will zur macht*.[88] However, before the narrator (and Nietzsche's *mensch*) can learn to love what he wills, he must undergo a process of unlearning.

'Unlearning'

Unlearning is a key shared element in the writings of Nietzsche and of Hinton's *An Unfinished Communication*. Nietzsche frequently mentioned unlearning as a positive capability; the words *verlernen* and *vergessen* often appear in 'On the Uses and Disadvantages of History for Life'. Both terms are commonly translated into English as 'to forget'; however, both Oscar Levy, who provided the first English translation of Nietzsche's complete *Werke*, and R. J. Hollingdale, who has recently translated Nietzsche for Cambridge University Press, differentiate between verlernen and vergessen by rendering the former as 'to unlearn' and the latter 'to forget'. To unlearn seems to be the more literal translation of verlernen. The use of the two different words implies a difference between ways of dealing with the past, between active transcendence and passive relinquishment.

To unlearn something is to actively engage with it in order to alter its effect on one's present perception, to treat the past as a palimpsest rather than a blank slate. In opposition to unlearning is forgetting, which implies an involuntary, unhistorical relationship with the past. For example, in the first edition of *Zarathustra*, Nietzsche implied that verlernen is a purposeful, willed activity: 'Verlernte er [the will] den Geist der Rache und alles Zähnefnirchen? [Can the Will unlearn the spirit of revenge and teeth-gnashing?]'.[89] Turning to *An Unfinished Communication*, we see that unlearning this 'spirit of revenge and teeth-gnashing', which is the modern individual's reaction to the past, plays a major role in the plot development.

The story is told as the first-person narration of a middle-aged man who is overwhelmed by the 'sordid details' of his personal history, and the history of Western civilization that is represented by the dilapidated New York City quarter he finds himself in, 'in which the streets, winding and squalid, with low buildings on either hand, resemble the alleys of an old world city'.[90] Walking down a street in this quarter, the narrator

---

86 Nietzsche, *The Birth of Tragedy*, 146.
87 Hinton, *An Unfinished Communication*, *Scientific Romances*, 110–11.
88 While the noun *macht* translates most directly as 'power', in its verb form *machen* translates as 'to make'.
89 Nietzsche, *Also Sprach Zarathustra*, 208. This text is an impression of the first edition.
90 Hinton, *An Unfinished Communication*, *Scientific Romances*, 109.

dimly registers the multitudes of 'notices of decaying trades' and 'the advertisements of professors in the last stage of indigence', until an unusual sign catches the corner of his eye. The sign reads: 'MR. SMITH, UNLEARNER'.[91] The narrator finds the concept of unlearning exciting, and the plot of the story hinges on his quest to find the Unlearner.

The narrator's reaction to the Unlearner's advertisement is worth quoting at length:

> How pleasant it would be to let pass away some of the verbiage I learnt at school – learnt because teachers must live, I suppose. The apeing [sic] and prolonged caw called grammar, the cackling of the human hen over the egg of language – I should like to unlearn grammar. The sense came over me [...] of how much I should owe to any man who would rid me of what I learned at college – that plastering over the face of nature, that series of tricks and devices whereby they teach a man knowing nothing of reality to talk of it as if he did. There passed before my mind that pallid series of ghosts, ghosts of what had once been one man's living, practical work, the books by which professors – because they must live, I suppose – keep younger men from life and work.
>
> A gleam of hope came over me that I might forget my philosophy lectures and the teaching of that bespectacled Doctor of all the sciences, who always turned the handle the wrong way, while he told us the principles by which things go.
>
> [...] If all these were to sink and disappear from me, then perhaps I should be face to face not with a spectre, not an instance and example of a phase, a formula, a barren set of words.[92]

At this point in the story, the narrator conflates unlearning and forgetting; it is only after his experience of eternal recurrence that he realizes the difference. There is also an important difference between wanting to slough off the tricks of language so as to come into contact with something 'real', and the realization that no such objective reality exists.

There is a sense – in both Hinton and Nietzsche – that the acceleration of the proliferation of knowledge and facts in the nineteenth century was crippling the spirit of youth and innovation. Like Hinton's narrator, in 'On the Uses and Disadvantages of History', Nietzsche derided the burgeoning of scientific discourse:

> Science has certainly been pushed forward at an astonishing speed over the past decades: but just look at the men of learning, the exhausted hens. [...] They can only cackle more than ever because they lay eggs more often: though the eggs, to be sure, have got smaller and smaller (though the books have got thicker and thicker).[93]

The younger men who are kept from 'life and work' by the burden of the past in Hinton's text are akin to Nietzsche's 'youth': 'that first generation of fighters and dragon-slayers' who must create space for the next generation through having 'unlearned many things and even [...] lost all desire to so much as glance at that which these cultivated people [of the present] want to know most of all'.[94] Although he does not yet understand it

---

91 Ibid., 109–10.
92 Ibid., 110–11.
93 Nietzsche, *Untimely Meditations*, 99.
94 Ibid., 121 and 122.

at the beginning of the story, Hinton's narrator must unlearn the desire to know the 'truth', to realize there is no singular truth and to move beyond mourning the loss of this ideal.

At the beginning of *An Unfinished Communication*, the narrator's quest to find the Unlearner is founded in a nihilistic wish to deny the past and therefore existence. When he first finds the Unlearner, the narrator asks for help in forgetting his personal history, and it is at this point that the Unlearner differentiates between unlearning and forgetting: 'Wherefore forget? What you have been is the food on which your soul lives. Think how closely connected memory and self-consciousness are; snap the last chord of recollection and you would lose the sense of personal identity'.[95] The narrator's wish to forget is a desire not for unlearning, but for forgetting which, when taken to its extreme, becomes an act of self-annihilation. Here the narrator is dying-to-not-know, as he confirms to the Unlearner: ' "I do not want these state moralities", I said; "we are fettered and bound by the past, and oblivion – utter oblivion – is a cheap price to pay for freedom" '.[96] His request to forget is thus a form of *reactive* nihilism, driven by the desire to wreak revenge on 'that which was' by denying its existence, and life. This revenge-denial is one extreme in the will's range of possible reactions to the failure to transcend a linear model of time. As the Unlearner explains, the narrator's nihilistic drive for oblivion – if successful – would place him outside the range of the human; snapping the connections between memory and consciousness would turn him into a sort of unhistorical animal.

Through his contact with the Unlearner and his later experience of eternal recurrence the narrator grows to recognize his potential for enacting a sort of 'active' nihilism, which is life-affirming. Rampley defines active nihilism as resting on 'the recognition of the perspectivism of interpretation, acceptance of the contingency of knowledge, and the recognition that "knowledge" is interpretative will to power'.[97] This re-cognition is only possible because he is able to unlearn the model of linear, irreversible time. The four-dimensional view of the eternal recurrence of his life allows the narrator to unlearn – which is itself a playful reinterpreting of knowledge – rather than simply to forget: here is the difference between re-writing and erasing. It is the creative will that reinterprets the past and thus, by implication, the present and future selves. Therefore, as Rampley observes, a 'crucial element' of Nietzsche's active nihilism is 'the absence of nostalgia for anything metaphysics might regard as "true knowledge" '.[98] It is through the thought-experiment of eternal recurrence that the subject learns to transcend the desire for absolute knowledge.

Eternal Recurrence in Nietzsche and Hinton

In the section of *Thus Spake Zarathustra* titled 'Of the Vision and the Riddle', Nietzsche's Zarathustra tells his audience of a dream in which he comes upon a gateway: 'From this

---

95 Hinton, *An Unfinished Communication, Scientific Romances*, 120.
96 Ibid., 120.
97 Rampley, *Nietzsche, Aesthetics and Modernity*, 219.
98 Ibid., 219.

Gateway called Moment a long, unending road runneth back – behind lieth all eternity'.⁹⁹ This road also stretches from the gateway in the opposite direction, representing the future. In his dream Zarathustra approaches this gateway while climbing up a steep mountain path. A dwarf, who represents the 'Spirit of Gravity', or the sense of the past, weighs him down on this climb.¹⁰⁰ Seeing the gateway and the two roads, Zarathustra questions the dwarf: 'Believest thou, Dwarf, that these roads controvert one another *eternally*?'. The dwarf responds by threatening Zarathustra's hopes of surmounting the past with the idea of eternal recurrence: 'Time itself is a circle'. If this is the case, then, asks Zarathustra, 'must not we all return eternally?'. Upon formulating this question, Zarathustra is horrified by this newly understood model of time: 'Thus spake I and ever more softly! For I feared mine own thoughts and the thoughts behind my thoughts'. In Zarathustra's image of eternal recurrence, there is no room for change:

> Must not all that *can* have run already run this road? [...] And if all hath already been [...] must not this Gateway also have previously existed? [...] For all that *can* run – even the length of this long road – *must* run it yet again.¹⁰¹

The emphasis on 'can' here highlights the problem of eternal recurrence *of the same*: if this is the case, then there is no possibility of creating anything new because all that could be created – the future – has *always* already existed and always will. This view of time is a suprahistorical view, and as such is a strong antidote to a linear, irreversible model of time and the obsession with regaining a lost past. However, Nietzsche noted that this suprahistorical view also precludes the possibility of any will to action just as much as a linear, irreversible model of time does.

Rampley argues that in using the idea of eternal recurrence as a thought-experiment, Nietzsche 'is suggesting that by viewing time cyclically, one might conceive of human agency and social activity differently' so as to transcend the will's ressentiment.¹⁰² However, Zarathustra's image of the present moment as a gateway moving along a circular track of time is problematic in that it similarly precludes access to the past. Robert Gooding-Williams explains this paradox:

> To suppose one's will were ensconced within an eternal, present moment would be to deny, in effect, that one's will could ever cease to will what it was willing at that moment, for whatever it was willing at that moment it would have to will eternally. [...] From the viewpoint of such a will human being-in-time would involve [...] no capacity to revalue and redeem passions of the sort that claimed generations past.¹⁰³

Thus the view from the gateway is unhistorical. Zarathustra challenges the discontinuity of (to borrow Hinton's words) this 'fugitive now': 'Are not all things thus knotted so fast

---

99 Nietzsche, *Thus Spake Zarathustra*, 142, emphasis removed.
100 Ibid., 140.
101 Ibid., 142, original emphasis.
102 Rampley, *Nietzsche, Aesthetics and Modernity*, 148.
103 Gooding-Williams, *Nietzsche's Dionysian Modernism*, 225.

together that this moment draweth after it *all* that is to come? *And therefore* itself also?'[104] The gateway-perspective is also an illusion: the present moment is never entirely divorced from the past or the future. As Rampley notes, 'What is implicit in Nietzsche's discussions of Eternal Recurrence is a questioning of time into heterogeneous "aspects"'.[105]

It is important in the gateway scene that Zarathustra's suprahistorical, spatialized vision of time allows him to understand the present moment as a site that is created by the past, that the mere existence of the present moment precludes any real possibility of denial of the past. Crucially, Rampley argues, such a perspective of time makes 'now' the place where

> the past is always being refigured, but its content never remains the same. *Amor fati* logically follows from this. Far from being a fatalistic acceptance of everything that has been, it is an affirmation of everything that has been in light of the recognition that the meaning of history can always change depending on the content of the present.[106]

Here the present moment is far from being the site of absolute denial of the past, of passive forgetting. Occurring here is an active unlearning of the metaphysical desire for transcendence of the past that would entail transcending materiality altogether. As Hinton wrote in 1888, 'to think of ourselves as any other than things in space and subject to material conditions, is absurd'.[107]

Turning to *An Unfinished Communication*, we can observe a vision of recurrence that is similar to Zarathustra's. The narrator seeks out the Unlearner in a remote coastal village in New England. After the conversation with the Unlearner quoted above, the Unlearner advises the narrator to visit an even more remote village along the coast. The narrator spends several days in this small fishing village until one afternoon, when he is out walking on the beach, he is caught in the incoming tide and drowns. While he is drowning, the narrator experiences recurrences of his life to date; however, after he has died, he begins to experience something very similar to Nietzsche's eternal recurrence. The scenes that comprise the narration after the death of the narrator have an uncanny, almost cinematic quality to them. Time, as measured by progressive movement forward, becomes distorted, revealing the relativity of perspectivism. The narrator is a child again, riding with his father on a streetcar:

> I was actually with my father, sitting by his side. Through a window I saw a horse's head jogging up and down. The horse was trotting; I could see the motion of every step – yet it was going backward all the while. I asked my father – we sitting in the street car together – 'Why does that horse go backwards when he is going forward?' But my father was reading his paper and did not answer me. […] The car stopped, and I saw with satisfaction the horse go on, dragging his load quite fast after him.[108]

---

104 Nietzsche, *Thus Spake Zarathustra*, 142, original emphasis.
105 Rampley, *Nietzsche, Aesthetics and Modernity*, 151.
106 Ibid., 152.
107 Hinton, *A New Era*, 94.
108 Hinton, *An Unfinished Communication*, *Scientific Romances*, 157.

Here Hinton seems to have drawn on the strange visual spectacle of film running in reverse to represent the modern sensation of speeding forward in a streetcar. The first documented example of reverse film projection occurred in Louis Lumiere's *Charcuterie Mecanique* (1895), though there are earlier examples of public motion picture exhibitions in the United States, such as Thomas Edison's Kinetoscope screenings in New York City beginning in April 1894, Ottomar Anschütz's électrotachyscope and Eadweard Muybridge's zoopraxiscope, both of which were displayed at the Chicago World's Fair in 1893.[109] It is possible that Hinton saw such an exhibition.[110]

Time is then collapsed into space during the eternal recurrence scenes and Hinton's four-dimensional visionary is able to view his entire history, depicted as abrupt, cinematic 'jump-cuts' from one time and place to another. First the narrator is a child in the streetcar with his father, then a young man in America, then a student at Cambridge, an adult in Vienna, then in Central Park, then in the squalid quarter of New York City, and finally, on the New England coast. Time is dissolved in this montage; because the narrator shifts abruptly from one place and time to another, all times appear to be the same time. He explains:

> Just as truly as I am lying here, rising and sinking with the heave of the waters, so I am in each of the scenes and places I have ever been in, living and acting in them. It has been coming over me in scenes which I thought were vivid memories, but now I know they are actual presences. I am a child with again other children. I see my father: each step, each act, each little thing, I go through again, living the very life I lived before. I am a man with other men and women. I am moving, speaking with them, and they with each other – not only I, but all are living as they ever lived.[111]

The narrator experiences each moment of his past as *presence*, not memory. In his attempt to represent this uncanny moment of collapsed temporality as the recurrence of personal historical events, Hinton reached outside traditional literary narrative toward filmic narrative. Later, at the height of literary modernism, we find T. S. Eliot collapsing temporality, through, as Joseph Frank explains, 'the juxtaposition of disparate historical images' such as 'Mr. Eugenides, the commercial traveller from Smyrna' in *The Waste Land*. Frank expands Alain Robbe-Grillet's claim that 'the cinema knows only a single grammatical modality; the present of the indicative', to include the work of Eliot, Joyce and Pound.[112] It is fitting that the narrator's montage of recurrence begins with the re-presentation

---

109 Wierzbicki, *Film Music*, 16.
110 That Hinton looked toward new media technologies for ways of conceptualizing the fourth dimension is demonstrated in the first series romance, 'A Picture of Our Universe' (1886), in which he compares the movement of the earth (and every individual on it) across the ether to the movement of a stylus over a phonographic cylinder. The grooves in the cylinder/ether determine the movement of the stylus/earth, but through an act of volition, the individual can alter their particular 'groove' in the ether.
111 Hinton, *An Unfinished Communication, Scientific Romances*, 173.
112 Frank, *The Idea*, 78.

of a childhood journey on a streetcar, where he re-lives his sense of wonder at the discovery of a new way of seeing. This is one of the 'perception-altering experiences' of nineteenth-century modern culture that Keith Williams identifies: 'New forms of spectatorship had already been made possible by the rapid succession of images produced by innovations in transport' as the 'mobilised and "framed" viewpoints through train, tram and later motor-car window compressed temporality, foreshortened distance and blurred landscapes [...]'.[113]

During this sequence of temporal jumps, the recurrence montage is interrupted by a vision that allows the narrator to unlearn his nostalgia for 'true' knowledge. He has a mystical vision of 'the spacious courts of heaven', where St Paul and St Simeon Stylites sit in judgement. They are approached by a strange penitent:

> A figure drew near with a mean robe flung all about her, and in her arms a great bundle. But her walk was as the walk of a humbled queen, and in her voice there was the rippling of the waters, the sighing of the winds, the song of the birds, as she prayed, 'Judge me, I have stolen these.'[...]
>
> She stooped, rested the bundle on heaven's floor, and opened it. The mean, worn covering was folded back, and there was nothing and yet everything – everything that men have seen of colour in the sunset or in the deep sky. There were the grace of the dappled limbs of the fawn, the lines of strength of the tiger, the wonderful green of the forests, the all-burying forests in their wonderful mazes, the delicate blue of the distance, the depths of the ocean, the semblance and likeness of everything there has been on earth. There, without the substance and body of them, were the grace and beauty of human countenances, the bloom on the cheeks, the vermeil lips, the glance of loving, passionate, ardent, alluring eyes, and the quiet, long, still gaze of dark eyes. There was the glamour and grace and beauty of all that man has ever loved to gaze upon [...]. There were the flash of white limbs through translucent water, the raised arms of Venus, her head waving like a flower between them. All was there; not the substance of things, but the show of them [...]. 'All these are not mine, and I have taken them, all the sounds of my voice, and these I have brought here. Henceforth I will be mute and without all these. Oh, judge me.'[114]

The mysterious figure is what poets and scientists have called 'Nature', but, as the woman explains:

> It was mine [...] at the creation to keep the busy atoms dancing, to turn and twist them on their moving course, playing the shuttle of vibrations in all the system. But men wove robes and garments, inventing light and colour, placing light and colour and sound before me. They praised me, calling me Nature and wonderful, beautiful. I, because I liked their praise, [...] I put on, feigning to be what men praised – I, who all the while have no part in any of these things, whose it is to move the atoms on their ceaseless wheeling.[115]

---

113 K. Williams, *H. G. Wells, Modernity and the Movies*, 13.
114 Hinton, *An Unfinished Communication, Scientific Romances*, 162–64.
115 Ibid., 164.

What humans call 'Nature' is, in fact – like Hinton's Persian king – a void. The words attached to it are the concepts through which the human consciousness encounters the external world.

After the figure abandons the shape given to her by humanity, she returns to earth where 'a new day began. The light of dawn, the sunset of even, no longer were what man put on her, but were of Nature's own'.[116] At this point in the text we may be tempted to conclude that, contrary to abandoning a metaphysical wish for unmediated contact with the thing-in-itself, the narrative is celebrating it. However, the narrator's rhapsodies about the 'new day' on earth are rudely interrupted: ' "Very pretty, white limbs flashing through the water is good," says a voice by my ear'. This is the voice of what the narrator describes alternately as 'my intimate friend' and 'this Pan creature', the earthy, sensual side of his own nature, which drags him back to the experience of embodiment.[117] The narrator observes that this creature is an integral part of every human: 'As I walk, I see one and another of this kind [his intimate friend] following or arm in arm with men like myself, and I know that this city is full of them'.

The narrator denies this aspect of his humanity three times in the vision, but the creature undermines the narrator's attempts to transcend himself, telling him that such attempts make 'yourself endlessly wretched [...] you know you have no joy in life save for me, and all you think or do is to give me pleasure'. The narrator challenges this claim that he is entirely a pleasure-led being; however, he also accepts the creature as part of himself, replying, ' "It is not so," I say. But he is close to me; he takes my arm familiarly; I know I shall never be rid of him and *do not want to*'.[118] In acknowledging and even embracing the physical (as opposed to metaphysical) side of himself, the narrator offers a damning indictment of the duplicity of his own culture's values, using the language of Victorian propriety:

> I know that this friend, this part of me, had fooled me; [...] when I thought my motives were so different. [...] This Pan creature [...] weaves the exhalations of the earth into the shapes he wills, [...] assumes the garb of *pity, duty, sacrifice*, speaks in the name of *utility, common sense, sanctity*, and whatever he finds will gain his ends.[119]

In acknowledging that it is 'this part of *myself*' that interprets and shapes the world as he wills it, the narrator recognizes 'the perspectivism of interpretation, [...] that "knowledge" is interpretative will to power'.[120] The values of Victorian culture, the morality by which he has alternately justified and condemned his past behaviour, are open to reinterpretation. Here the narrator moves beyond good and evil, in a Nietzschean, 'extra-moral sense', as opposed to the 'moral person [who] selects what they require for their own self-preservation and fails to see that in the general economy of the whole [...] life is the

---

116   Ibid., 165.
117   Ibid., 166.
118   Ibid., 167, emphasis added.
119   Ibid., 170, emphasis added.
120   Rampley, *Nietzsche, Aesthetics and Modernity*, 219.

unity of pleasure and pain, of joy and suffering, of good and evil'.[121] Like the student in 'The Persian King', he has become a 'true personality'.

Having gained a 'higher' perspective on his life and unlearned the desire for metaphysical transcendence, the narrator's vision of eternal recurrence expands even further. The visions of recurrence of past events from his life have allowed him to unlearn the linear, irreversible model of time and the revenge-driven nihilism that results from such a model: 'Yet though the bond – the fetter of unalterability – is on me, all feeling of loss and of the irrevocable passes away'.[122] Here Hinton's dramatization of the experience of eternal recurrence functions the same way as Maxwell's daemonic thought-experiment: both illustrate that the limits we perceive in 'objective' reality are actually the limits of epistemology.

With his modern sense of ressentiment removed from him, Hinton's narrator's perspective alters even further: 'But watching closely with so eager a curiosity, I see that each of us [in the recurrences of his life] is not doing exactly the same – and see, our lives are altering'. This is not eternal recurrence of the *same*, but eternal recurrence with variation. Through an act of creative will, the mind is able to continually reinterpret the past in light of the present: 'I feel that sudden touch with Nature lays on all those that die, saying to them "Know! I am ever changing, altering. With me everything is in a state and stage of development. I allow not anything to be cast in a rigid mould, not even thy past life *in thy imagination*"'.[123] Thus the past is always open to appropriation by the creative consciousness. The will cannot change 'it was' by travelling back along the line of time, but time as perceived from the fourth dimension is always already completely present in each moment. At this point in the text the narrator accesses his four-dimensional aesthetic will, which is

> the will that acts along the whole line [of time], the will whose body is the whole life – that I catch, fragmentarily present here and there in my life – that will, shown, not in great things, but in minute, almost invisible changes, that will is what I prize and treasure, for it is the means whereby my life alters, the means by which it is what it has become.[124]

Here Hinton dramatized what, in Ansell-Pearson's words, 'Nietzsche is demanding of his readers[, which] is nothing less than that they give birth to themselves'.[125] The narrator learns to become what he is by 'plunging in [his life] part by part' and willing the four-dimensional whole. The communication to the reader must remain unfinished here because this is a project with no end.

In the second text of the second series, 'Many Dimensions', Hinton acknowledged that there may be an infinite number of dimensions beyond a fourth, and throughout the *Scientific Romances* we observe polysemy and contradiction, both of which resist the

---

121 Ansell-Pearson, 'Who Is the *Übermensch*?', 320.
122 Hinton, *An Unfinished Communication, Scientific Romances*, 174.
123 Ibid., 174, emphasis added.
124 Ibid., 174–75.
125 Ansell-Pearson, 'Who Is the *Übermensch*?', 326.

closure of absolute meaning. Hinton recognized this himself in his preface to the 1895 edition of *Stella and An Unfinished Communication* when he wrote that, in this book 'one line, one feature of the landscape of the land to which these thoughts lead, and only one, has been touched upon. But there are many, and each explorer would probably select a different one'.[126]

Speaking in Victorian terms of gendered, masculinist temporal logic, Hinton, like Nietzsche, privileged a 'feminine' aesthetic. Zarathustra, whose 'quest for truth and meaning culminates in the recognition of "eternity" understood as a woman', parallels the experiment in *Stella*, where Hinton chose a female subject to experience eternity.[127] In his 'Prelude to a Philosophy of the Future', Nietzsche presented us with a miniature thought-experiment: 'Supposing truth to be a woman – what?'. Unfortunately, Nietzsche continued, philosophers, 'when they have been dogmatists, have had little understanding of women'.[128] This is because, as Ansell-Pearson explains, for Nietzsche woman was 'like truth – plural, polysemous, a dissimulating veil behind which lies not *the* truth but another veil, another mask'.[129] Hinton's 'Nature' in *An Unfinished Communication* sheds her conceptual masks […] to become what? There is no language to describe it and, as Hinton's narrator realizes, it is in the process of describing that pleasure and pain that life is, in fact, created.

---

126 Hinton, *Stella and An Unfinished Communication: Studies of the Unseen* (1895), n.p.
127 Ibid., 327.
128 Nietzsche, *Beyond Good and Evil*, 31.
129 Ansell-Pearson, 'Who Is the *Übermensch*?', 327.

Part II

READING *THROUGH* THE FOURTH DIMENSION

Chapter Four

FOUR-DIMENSIONAL CONSCIOUSNESS: THE CORRESPONDENCE BETWEEN WILLIAM JAMES AND CHARLES HOWARD HINTON

In my analysis of the second series of *Scientific Romances* we saw how Hinton began to explore the significance of four-dimensional consciousness for the individual subject. In particular, the narrator in *An Unfinished Communication* experiences four-dimensional consciousness as something that is continually developing and changing over time. Although it was latent in the first series of *Scientific Romances*, in the second series and the years that followed, the model Hinton was developing positions consciousness as a continual *process* of realization of higher and higher dimensions of being. He made this view explicit in a letter to William James in 1897:

> I'm sure I much agree with you about the variegated nature of existence. In fact all attempts to run a principle thro' things seem to me to be like plastering a piece over the whole. Unification is a means. Because it is useful in science it gets to be assumed to be an end in itself[.] A more and more complex whole ever revealing itself in unusual forms – that seems to me the end of philosophers, with theories and principles, the arid analysis, merely to be tolerated because they enable us to carry the mass of details necessary to be grasped before we can apprehend the higher reality[,] the higher personality[,] the actual being.[1]

It is not surprising, then, that Hinton was interested in James's work. James was also trying to construct a new model of consciousness and he was aware of the limitations of using language to describe a process of consciousness that would always exceed itself. He called his model of consciousness 'conjunctive' (it 'flows' in a 'stream' as opposed to moving in discontinuous 'units' like a 'train'): it was a 'reinstatement of the vague and inarticulate to its proper place in our mental life'.[2] Here James was foregrounding the transitory, 'dumb or anonymous psychic states' that contemporary psychological theory had suppressed.[3] To focus on these transitory states is to undertake a dynamic and creative project, because 'every smallest state of consciousness, concretely taken, overflows

---

1 Hinton to W. James, 10 July 1897, 1–2. All letters from Hinton to James cited here are from the Letters from Various Correspondence section of the William James collection at Houghton Library, Harvard University, bMS Am1092.
2 W. James, *Writings, 1902–1910*, 164.
3 W. James, *The Principles of Psychology*, 1: 246.

its own definition'.⁴ As soon as we think we have captured the meaning of an experience, it expands beyond our conceptualization; this perpetual sense of 'more' underpinned much of James's writing, and James's psychological theory of consciousness naturally fed into his later pragmatist and radical empiricist writings. In these later writings, James theorized that consciousness was actually *created* through a process of ambulatory experience. While numerous critics have observed the importance of James's work for the rise of literary modernism, none have acknowledged the correspondence between his ideas and those expressed in Hinton's hyperspace philosophy, and none have considered the fourth dimension within the context of the contemporary development of pragmatism as a philosophical and aesthetic approach. Hinton's various attempts to guide his readers through the process of imaginatively constructing the fourth dimension can be read as a similar project of reinstating the vague through the metaphorics of fluidity and movement.

Writing of William James's project of 'the reinstatement of the vague', Joan Richardson observes how the 'development of what was laid out in [...] *Principles* is traced into its branchings in *The Varieties of Religious Experience* to "re-crystalize" in Pragmatism'.⁵ Thus, James's own ambulations through psychology, spiritualism and pragmatism share a common root. Richardson continues on to highlight James's fluency in European languages, noting that we should consider 'his central and extensively repeated use of vague – "wave" in French' within the context of contemporary scientific investigations into the dynamics of light, sound, electricity and magnetism.⁶ Richardson is right to observe the overlapping interest in *movement* here, though I argue that in their shared participation in what Jonathan Levin has aptly described as a 'poetics of transition', these philosophers and scientists were expressing a larger cultural trend that would be most explicitly embodied in twentieth-century modernist literature and art.⁷ What is needed in an examination of this 'poetics of transition' is not a 'history of ideas' model of influence, but an ambulatory, ' "history of consciousness" method [... that] arises in turn from the assumption that a Cartesian dualism of subject-object disjunction and division is not the actual state of affairs – in cognition, perception, or indeed any of the fundamental relationships which comprise the human condition'.⁸ The story of Hinton's hyperspace philosophy is not one of background and foreground or one-way influence. In the present chapter I read Hinton *beside* the psychological and philosophical writings of William James, observing how both men drew on shared tropes of movement, fluidity and process to describe what they both ultimately decided was indescribable within the current limitations of language: the sense of something 'more', an experience of being that exceeds contemporary models of consciousness. Both James and Hinton drew on the same metaphor – Gustav Theodor Fechner's 'wave scheme' or 'mother-sea' of consciousness – to express their sense of this other dimension of being. I examine their use

---

4   W. James, *Writings, 1902–1910*, 760.
5   Richardson, *A Natural History*, 17.
6   Ibid., 108.
7   Levin, *The Poetics of Transition*.
8   Hocks, *Henry James and Pragmatistic Thought*, 43–44.

of the mother-sea metaphor in their separate writings and their shared interest in the concepts of volition and attentiveness as means of accessing a higher aesthetic sensibility. Because I am breaking new ground in reading James's work beside Hinton's and considering his more familiar stream-of-consciousness theory *through* the unfamiliar lens of the fourth dimension, I begin this chapter by offering a more concrete foundation for my consideration of the correspondence between William James and Hinton, by analyzing the actual written correspondence (housed in the William James Collection at Harvard University) between these two men.

The Hinton–James Correspondence

In true Jamesian fashion, I intend to take advantage of the ambiguity of the word 'correspondence' here. Before establishing a case for the 'accidental, the partial and the metaphorical' moments in which the ideas of William James and Hinton overlapped within their published writings – the ways in which their ideas and their work *correspond* – I will contextualize this analysis by examining what remains of the actual correspondence between the two men. This remnant consists of eleven letters from Hinton to James, dating from 1892 to 1907. The relationship between Hinton and James has never been fully explored, and if we are to understand how the ideas of one reinforce (and, at times, seemingly, respond to) the other's, then it is important to consider to what extent these two men may have interacted with each other.

Shortly after Hinton arrived in the United States from Japan, he wrote to William James:

> As soon as I landed in America or very soon after I heard that you had gone abroad. I was very sorry as I had looked forward to having a talk with you.
>
> The whole 4 di. theory has turned right round in my mind. For the geometry of the thing it is right to imagine ourselves indefinitely flat in the 4th dimension. But that there are in nature no two dimensional beings shows that there being assumed higher space we must be higher space beings in a higher space world in contact with higher space existences. [...] Now that means that our limitation must be one of consciousness – the totality of space organs working under this condition – of perceiving in 3 space. And the only way in which the higher space existence can be seen is in the change of our existing objects. Assuming this it is a question of how much change is due to a moving consciousness in a higher space world. We ordinarily assume consciousness – indefinitely extended space to be fixed & things to move about in it. But here is the possibility of a different view [of] permanent things and a moving consciousness. I have developed the idea a little & will send you something about it. [...] I can do the explanation of the whole subject infinitely better now the new models are as simple as *a b c d*. [...] Being over here is a rediscovery of mind in myself and others. I am glad you liked the Japanese picture.[9]

Though this is the first extant letter from Hinton to James, the tone, in addition to its references to the cube exercises ('the new models') and 'the Japanese picture', as well as

---

9   Hinton to W. James, 5 October 1892, 1–2.

the assumption of James's prior knowledge of his '4 di. theory', indicate that Hinton and James had been in contact before 1892. How or when Hinton and James first met is unclear; James was a friend and admirer of Shadworth Hodgson, a British philosopher and forerunner of pragmatism, and he cited Hodgson with approval frequently throughout his writings. Hodgson was also a friend and admirer of James Hinton, and he wrote the introduction to James Hinton's posthumous *Chapters on the Art of Thinking*, which the younger Hinton edited. That James was also an admirer of James Hinton is evinced by his approving references to the elder Hinton in his published writings.[10] In his unpublished biography of Charles Howard Hinton, Marvin H. Ballard speculates that Hinton and William James may have met through Hodgson, when James was travelling in Europe in 1882 and 1883.[11] While this is certainly a possibility, there were in fact numerous overlaps in Hinton and James's circles. Hinton lived and travelled in Japan from around 1887 to 1892; during this time numerous Boston intellectuals, including recent Harvard graduates William Sturgis Bigelow and Ernest Fenollosa, were in Japan. With such a small number of Westerners in the country at the time, it is likely that Hinton met and perhaps even befriended some of these men.[12]

Whenever or however they met, it is clear from the remaining correspondence that James was kindly disposed toward Hinton, and the two men had a friendly relationship. There are references to them meeting in person on at least during two different occasions during the period of correspondence. Hinton remarked that 'seeing you was pleasant as it has awoke a great many unprofitable thoughts. It was so strange that a professional being should have any interest in what I do'.[13] Hinton managed to obtain a mathematics lectureship at Princeton in 1893, but he was unable to keep this post, and after his contract was terminated in 1897 he took an assistant professorship at the University of Minnesota. By 1901 he was on the move again, taking a post at the Royal Naval Observatory in Washington, DC. Hinton appeared to have struggled to gain academic credibility in the United States, which is indicated in his comment to James above, about 'professional being[s]'.

Hinton's aversion to institutionalized intellectuals increased over the period of his correspondence with James, as demonstrated in a letter from William James to F. C. S. Schiller in 1904, when James remarked with amusement and approval that 'in a letter from C. H. Hinton yesterday, he says: "The academic mind secretes thought and contempt together, in about equal proportions!"'.[14] James was known for his broadmindedness and interest in subjects on the fringe of academic respectability, and he appears to have offered Hinton moral and intellectual support. Shortly after Hinton's

---

10 See, for example, W. James, *Writings, 1878–1899*, 532.
11 Ballard, 'The Life and Thought', 49.
12 See Lears, *No Place of Grace*. Murray's *A Fantastic Journey* also demonstrates the interconnectedness of the small community of Westerners in Japan at the end of the nineteenth century. For example, Hearn, who was close friends with Fenollosa, also knew Hinton, and stayed with him and his family for a time in Yokohama.
13 Hinton to W. James, 30 October 1896, 1.
14 W. James to Schiller, 2 September 1904. In Perry, *The Thought and Character*, 2: 505.

death his wife, Mary Boole Hinton, wrote to James, thanking him for 'the way in which you *stood by him* while he lived'.[15] In addition to James, Hodgson and Schiller, Hinton was apparently acquainted with another key figure of pragmatist philosophy, John Dewey. In the same letter to James, Mary Boole Hinton noted:

> My Princeton boy [Sebastian Hinton] has just left after his Thanksgiving visit. He has much of his father's ability together with more power of taking care of himself [...]. Prof. and Mrs Dewey should tell you about him. They have been good friends to me.[16]

As the archival evidence suggests that Hinton was circulating within the Anglo-American intellectual mainstream at the turn of the century, it is not surprising that he and William James met and exchanged ideas.

James apparently encouraged Hinton to publicize his work. In an 1895 letter, Hinton wrote: 'I have left your last letter of suggestion unanswered for so long because the offering on my part to give such a course of lectures as you suggest would be an act of presumption'.[17] He continued, asking James to 'suggest this subject to Mr Lowell', referring to what he would have liked to lecture upon: 'The axioms of geometry what they are and their significance in an epistemological point of view'.[18] Hinton's response indicates that James suggested Hinton apply to give a series of lectures at the Lowell Institute at Harvard. Hinton later would regret his decision not to pursue these lectures: in another letter he wrote that 'the Lowell Lectures would have been just the thing for me'.[19]

Hinton's refusal to undertake the Lowell Lectures, even though this was apparently an opportunity he desired, is puzzling. In the same letter cited above, he claimed that he had had neither the time nor money to attempt them. Ballard speculates that Hinton may have wished to avoid intense public scrutiny due to his bigamy conviction.[20] This is a very real possibility, as a similar personal scandal destroyed Charles Sanders Peirce's career at Johns Hopkins University in 1883. William James supported and encouraged Peirce personally and financially, helping him find lecturing work on occasion.[21] James apparently extended financial assistance to Hinton as well. In 1898 Hinton wrote to James:

> I find that I can just about make both ends meet here. But we have had a bad spell of typhoid with two of the boys & a good deal of other sickness. Happily all are well now. But I want to

---

15 M. B. Hinton to W. James, 2 December 1907, 1, original emphasis.
16 Ibid., 1.
17 Hinton to W. James, 19 October 1895, 1.
18 Ibid., 1.
19 Hinton to W. James, 30 October 1896, 2.
20 Ballard, 'The Life and Thought', 60.
21 See Brent, *Charles Sanders Peirce*. Peirce and Hinton had a likely mutual acquaintance and colleague in Simon Newcomb. Newcomb was a professor of mathematics and astronomy at Johns Hopkins with Peirce; there was an intense professional rivalry between Peirce and Newcomb, and it was Newcomb who brought Peirce's extramarital affair to the attention of the president

try to get something to do this summer. If you hear of anything you'll let me know. Under the circumstances you see for the present your loan must remain a gift.[22]

Hinton's bigamy scandal obviously changed his life profoundly, both personally and professionally. His life became one of frequent transitions: from England to Japan and then around various cities in the United States. This, too, may have impacted his hyperspace philosophy, which he increasingly expressed in the language of transition and movement.

As I demonstrate later in this chapter, both Hinton and James shared an interest in moments of contact with what Levin, by happy coincidence, calls 'another dimension of being'.[23] The letters from Hinton to James demonstrate that such instances were in fact a topic of discussion between the two men. In one letter – in an amused and sceptical tone – Hinton related the story of how one of his friends, a Mr White, was hypnotized by an acquaintance who had been inspired by a story that Hinton presented to a philosophical society in Princeton. Hinton had suggested, in a story that sounds similar to *An Unfinished Communication*, that

> just as in the body are the records of all the past physical planes through which the organism has gone in its development so in the consciousness are traces of all the past conscious life which lies back in a line from that consciousness. And by right suggestion it could be recalled. This of course was pure assumption – a postulate of a microphone of consciousness reaching back to the ur consciousness. But the man took it seriously and set to work to take [Mr White] back to the geologic antecedents of his conscious being, or to the primordial cell. Well you might expect they didn't get him back beyond his babyhood.[24]

However, they did 'regress' the man to his childhood, where he was unable to speak until they told him that 'he was a child in everything except faculty of speech'. After relaying a few of the reported memories of Mr White, Hinton asked James: 'Now do you think there is anything to be got as to the first impressions of childhood in this manner? Would this give a true means of observing a child's mind or would it merely be Mr White's present ideas of a what a child could think & feel?'.[25] Here the veracity of memory is questioned: Is it ever possible for an individual to access the same thought twice? Not according to James and Hinton, unless the individual is somehow able to transcend the normal horizon of attention to access something superior. Whether or not this is possible

---

at Johns Hopkins. Newcomb was later the director of the Nautical Almanac Office at the US Naval Observatory, retiring just before Hinton began work there. Newcomb also published work on four-dimensional geometry, and was active in the Washington, DC, intellectual community. Newcomb and Hinton's paths would have likely crossed often in Washington, and this would have only contributed to Hinton's reluctance to draw close scrutiny to his past.

22 Hinton to W. James, 12 August 1898, 3.
23 Levin, *The Poetics of Transition*, 65.
24 Hinton to W. James, 19 October 1895, 4.
25 Ibid., 5.

and, if so, how it could be attained, are problems to which both Hinton and James devoted most of their professional lives.

The conversation between James and Hinton was also informed by their readings of each other's work. Hinton noted reading James's collection, *The Will to Believe and Other Essays*, writing that 'I found [...] your book so full of ésprit so fascinating I've brought it to the seaside to read and hasten to send my acknowledgements'.[26] James read Hinton as well; in another letter Hinton wrote to thank James for his criticism, promising to 'embody the thing' commented upon in a longer piece of writing.[27] This letter, written in 1898, was found in James's 1904 edition of Hinton's *The Fourth Dimension*, suggesting that James may have referred back to the topic previously under discussion upon reading Hinton's later publication.[28] James also owned copies of the 1895 edition of *Stella and An Unfinished Communication* and Hinton's 1906 pamphlet, 'A Language of Space'.[29]

While there is no evidence that James ever supported or believed in Hinton's theory of the fourth dimension, he immediately recognized the ethical significance of his hyperspace philosophy. W. E. B. Du Bois's notebook from James's 1889 'Philosophy 4' course at Harvard indicates that James used Hinton's first romance, 'What Is the Fourth Dimension?' in his lectures on ethics. In his analysis of Du Bois's notebooks, Shamoon Zamir observes:

> James picked up the relativist and ethical implications of Hinton's argument, suggesting that 'we live in a 4th moral dimension separating us from animals' *(P4)*. This fourth moral dimension is the realm in which the will to believe negotiates its defence of a conditional, 'common sense' God against the 'absolute' God of 'speculative' theories such as Martineau's *(P4)*.[30]

---

26 Hinton to W. James, 10 July 1897, 1. Hinton did not refer to this text by name anywhere in the letter; however, I am confident in assuming that it was *The Will to Believe*. Firstly, the timing is appropriate. Also in the letter, Hinton mentioned details that correspond to *The Will to Believe*, the most explicit of which is Hinton's reference to 'The Brockton murder'. See also W. James, *Writings, 1878–1899*, 577.
27 Hinton to W. James, 12 August 1898, 1.
28 A pencilled note in the top margin of the first page of the 12 August 1898 letter states: 'Formerly laid in in [*sic*] James's copy of Hinton's The Fourth Dimension'. That James owned and read *The Fourth Dimension* is supported by Perry's claim that a marked copy of the book was sold from James's personal library. Perry cited in *The Will to Believe and Other Essays*, ed. Burkhardt, Bowers and Skrupskelis, 258, n. 22.34.
29 Harvard University's copy of the 1895 edition of *Stella and An Unfinished Communication* was previously owned by James, and a copy of Hinton's 1906 pamphlet, 'A Language of Space', is included in the William James collection at Harvard cited above.
30 Zamir, *Dark Voices*, 50. I have reproduced Zamir's citation, '*(P4)*' within this excerpt to emphasize that the quotations contained therein come from Du Bois's notebook, and therefore, presumably, James's lecture. James used the fourth dimension and the dimensional analogy in other lectures as well, including Harvard and public lectures. See James's lecture notes in W. James, *Manuscript Lectures*. Du Bois used the fourth dimension in an unfinished short story, 'A Vacation Unique', where a white Harvard student's skin is temporarily changed to black, so that he can experience 'the fourth dimension of race'. See Zamir's discussion of this manuscript fragment, 23–67; and Bentley, 'The Fourth Dimension: Kinlessness and African American Narrative'.

James's identification of Hinton's fourth dimension with his own theory of 'the will to believe' is telling. In the first series of *Scientific Romances*, Hinton's hyperspace philosophy was still very much grounded in the paradoxical tension between empiricist and idealist strains of Victorian scientific epistemology; he was particularly concerned with whether or not the fourth dimension is a 'real' space or epistemological effect. As Hinton's interest shifted towards the psychological fourth dimension, rather than looking for a space somewhere 'out there' he turned inward, away from 'permanent things' to the 'moving consciousness' of the individual subject. For Hinton, it was in the conjunctive 'moments' of this moving consciousness – the 'fugitive nows' of three-dimensional experience – that the fourth dimension is accessible.

The Fourth Dimension as Conjunctive Relations

In William James's famous 'Stream of Thought' chapter of *The Principles of Psychology* (1890), he constructed a model that challenged the two psychological conceptions of consciousness prevalent at the end of the nineteenth century, which can be divided roughly into empiricist and idealist camps. The empiricist model, which James called 'sensationalist', is a Locke-derived presentation of consciousness as a series of disjointed sensations linked through association. The idealist model, which James called 'intellectualist', claimed that consciousness is composed of discrete sensations that are linked together by transcendent ideas. For anyone who did not believe in a fourth dimension, Hinton's idea of four-dimensional consciousness would appear to be a variation of the idealist, intellectualist model of consciousness. However, for Hinton, who believed in a 'real' four-dimensional existence, James's theory of consciousness as 'flowing' had important implications – as it would for anyone attempting to access the four-dimensional imagination.

In his theory, James claims that

> consciousness [...] does not appear to itself chopped up in bits. Such words as 'chain' or 'train' do not describe it fitly as it presents itself in the first instance. It is nothing jointed; it flows. The images of the 'river' or the 'stream' are the metaphors by which it is most naturally described. In talking of it hereafter, let us call it the stream of thought, of consciousness, or of subjective life.[31]

It is partially the habit of language that has led to a misapprehension of the flowing 'structure' of consciousness. In fact, for James, consciousness lacks structure in any traditional sense. Transitions between 'states' of consciousness are actually states of consciousness as well:

> The transition between the thought of one object and the thought of another is no more a break in the *thought* than a joint in a bamboo is a break in the wood. It is a part of the *consciousness* as much as the joint is a part of the *bamboo*.[32]

---

31 W. James, *The Principles of Psychology*, 1: 239.
32 Ibid., 240, original emphasis.

Thus, according to James, both sensationalists and intellectualists are incorrect. Relations between different 'states' of consciousness really stop nowhere, 'and no existing language is capable of doing justice to all their shades'.[33] Language fixes consciousness into ideas; it shapes what the senses receive as percepts into substantives, or concepts. By focusing purely on substantives, one overlooks the sensations of conjunctive relations between 'states' of consciousness. However, according to James:

> If there be such things as feelings at all, *then so surely as relations between objects exist in rerum naturâ, so surely, and more surely, do feelings exist to which these relations are known.* There is not a conjunction or a preposition, and hardly an adverbial phrase, syntactic form, or inflection of voice, in human speech, that does not express some shading or other of relation which we at some moment actually feel to exist between the larger objects of our thought. If we speak objectively, it is the real relations that appear revealed; if we speak subjectively, it is the stream of consciousness that matches each of them by an inward coloring of its own. [...]
>
> We ought to say a feeling of *and*, a feeling of *if*, a feeling of *but*, and a feeling of *by*, quite as readily as we say a feeling of *blue* or a feeling of *cold*. Yet we do not: so inveterate has our habit become of recognizing the existence of the substantive parts alone, that language almost refuses to lend itself to any other use.[34]

The movement of consciousness, the process of transition whereby the mind comes to rest on a particular idea, is just as important (if not more so) than the ideas themselves. The problem is that although we '*ought* to say', we do not: language simply does not allow us to articulate or 'catch' the feeling of transition, because to capture it would be to destroy its transitory nature. Similarly, in 1906 Hinton observed: 'when we come to the problem of what goes on in the minute, and apply ourselves to the mechanism of the minute, we find our habitual conceptions inadequate'.[35] It is because of this habitual limitation that Hinton was only able to express his idea of a fourth dimension through a process of movement, through the 'Arabic method' of textual supplementation within the *Scientific Romances* and the study of space relations within his cube exercises. According to Hinton, it is because 'space sense' is the 'positive means by which the mind grasps its experience' that we must push *through* the limitations of spatial perception and language to access the four-dimensional consciousness.[36]

When introducing the 'practical work' included in *A New Era of Thought* – the extensive cube exercises that make up the second half of this text – Hinton explained:

> To begin it, we take up those details of position and relation which are generally relegated to symbolism or unconscious apprehension, and bring these waste products of thought into the central position of the laboratory of the mind. We turn all our attention on the most simple and obvious details of our every-day experience, and thence we build up a conception of the fundamental facts of position and arrangement in a higher world.[37]

---

33 Ibid., 245.
34 Ibid., 245–46, original emphasis.
35 Hinton, 'A Language of Space', *The Fourth Dimension*, 257.
36 Hinton, *A New Era*, 2.
37 Ibid., 97.

It is through examining these 'waste products of thought' – the overlooked sensations of spatial relations figured by Hinton elsewhere as the experience of duration – that the fourth dimension can be apprehended. That James's interest in the conjunctive relations of consciousness also developed out of his study of spatial relations is apparent in his 1907 discussion of his ambulatory methodology, where he reflected,

> I well remember the sudden relief it gave me to perceive one day that *space*-relations at any rate were homogeneous with the terms between which they mediated. The terms were spaces, and the relations were other intervening spaces.[38]

Previously, when researching spatial perception for his *Principles of Psychology*, James struggled with the subject; he found it tedious and difficult. In an 1879 letter to G. Stanley Hall, James wrote, 'I am composing a chapter on space for my psychology and find I have to re-read about all I ever read on that driest of subjects', and in a letter to Carl Stumpf eight years later he complained that 'space is really a direfully difficult subject!'.[39] His realization that space relations were spatial in themselves occurred while he was writing the *Principles*, because in this text he challenged the 'Platonizing school in psychology' which, he argued, claims that '*position*, for example can never be sensation'.[40] Conversely, for James:

> *Rightness and leftness, upness and downness, are again pure sensations* differing specifically from each other, and generically from everything else. Like all sensations, they can only be indicated, not described. If we take a cube and label one side *top*, another *bottom*, a third *front*, and a fourth *back*, there remains no form of words by which we can describe to another person which of the remaining sides is *right* and which is *left*. We can only point and say *here* is right and *there* is left, just as we should say *this* is red and *that* blue. [...] Thus it appears indubitable that all space-relations except those of magnitude are nothing more or less than pure sensational objects.[41]

James agreed with Kant in the assumption that humans immediately intuit space and spatial relations, but he wrote in a footnote that Kant 'is wrong, however, in invoking relation to extrinsic total space as essential to the existence of these contrasts in figures. Relation to our own body is enough'.[42] It seems then that James would have disagreed with Hinton's attempts to 'cast out the self' as an act of intuiting the 'absolute relations' of space without reference to the three-dimensional human body. Both Hinton and James, however, worked from the same premise in assuming that spatial relations *are*

---

38 W. James, *Writings, 1902–1910*, 898, original emphasis.
39 Perry, *The Thought and Character*, 1:16 and 70.
40 Jubin offers an illuminating, though brief, examination of William James's 1879 essay 'The Spatial Quale' as documenting James's changing thoughts about space during this period. See Jubin, 'The Spatial Quale'.
41 W. James, *The Principles of Psychology*, 2: 149 and 150–51, original emphasis.
42 Ibid., 151.

sensations, and are nearly inexplicable outside of the reference point of the human body. In 'Casting Out the Self', Hinton wrote:

> If we suppose that we are putting up the cubes in one room while another person is putting up cubes in an adjoining room; if we can tell him what we are doing, using the words right and left, he will be able to put a block exactly like ours. But if we do not allow ourselves to use the words right and left, but speak to the other person as if he were simply an intelligence without having the same kind of bodily organization as ourselves, we should find that, supposing he could put up the block of cubes, it would be a mere matter of chance whether he had put up the block as we had put it, or whether he had put it up in an image [i.e., mirror] way.[43]

There is no means of communicating the sensation of rightness and leftness without reference to the human body. Without the ability to gesture which is left and which is right, it is just as likely that one side of the cube (or block of cubes in Hinton's example) would be construed as left *or* right. In 'Casting Out the Self', Hinton strove to get *beyond* the sensation of leftness or rightness, because according to him, this three-dimensional limitation prevents us from perceiving the fourth dimension.

In 'Casting Out the Self' (1886), Hinton was still tentative about the likelihood that humans were four-dimensional beings themselves: 'Now *if* there are beings who live in a four-dimensional world, they must feel as habituated to it as we do to ours'.[44] Although as early as 'What Is the Fourth Dimension?' (1880), Hinton hypothesized that humans could be four-dimensional, in 'Casting Out the Self' he stepped back, suggesting that four-dimensional beings might exist but clearly differentiating them from humans. Hinton speculated that the minds of these beings were so much more developed than human consciousness that a human adult – with great effort – *might* be able to understand the thought processes of a four-dimensional child. The limitations of three-dimensional bodily sensations are something to be overcome here – hence, the need to 'cast out' the three-dimensional self in order to be able to perceive in four dimensions. In fact, in 'Casting Out the Self' Hinton concluded that the three expressions, 'Casting out the self – "Seeing as a higher child" – and thirdly, "Acquiring an intuitive knowledge of four-dimensional space"' are interchangeable.[45] Two years later, in *A New Era of Thought*, Hinton began to argue consistently in favour of the material existence of a fourth dimension of space and human four-dimensionality, and it was only after his arrival in the United States in 1892 that he began self-consciously approaching the task of 'getting sensations of the higher' space through the experience of duration. By 1904 he proclaimed with certainty that 'all attempts to visualise a fourth dimension are futile. It must be connected with a time experience in three space'.[46]

---

43 Hinton, 'Casting Out', *Scientific Romances*, 220.
44 Ibid., 224, emphasis added.
45 Ibid., 227.
46 Hinton, *The Fourth Dimension*, 207.

'A Moving Consciousness' in William James and Hinton

Earlier in this chapter I cited an extract from Hinton's first letter to James from 1892, doing so in order to examine the nature of their relationship. I return to this extract to consider how Hinton's idea of the fourth dimension had 'turned right round' in his mind:

> For the geometry of the thing it is right to imagine ourselves indefinitely flat in the 4th dimension. But that there are in nature no two dimensional beings shows that there being assumed higher space we must be higher space beings in a higher space world in contact with higher space existences. [...] Now that means that our limitation must be one of consciousness – the totality of space organs working under this condition – of perceiving in 3 space. And the only way in which the higher space existence can be seen is in the change of our existing objects. Assuming this it is a question of how much change is due to a moving consciousness in a higher space world. We ordinarily assume consciousness – indefinitely extended space to be fixed & things to move about in it. But here is the possibility of a different view [of] permanent things and a moving consciousness.[47]

Two statements in particular are of interest here: firstly, by this point Hinton was comfortable with assuming the physical existence of higher space ('there being *assumed* higher space'). Second, I highlight his proposal of 'permanent things and a moving consciousness', because this idea is where the works of James and Hinton overlap in surprising and productive ways.

Hinton's satisfactory demonstration (to himself, at least) of the material reality of the fourth dimension freed him to explore the psychological ramifications of his theory. In *An Unfinished Communication* we saw a dramatization of how one character experiences and is changed by 'the possibility of a different view of things [of] permanent things and a moving consciousness'. Hinton offered us a graphic representation of this same idea in his next book-length, non-fictional treatment of the subject, *The Fourth Dimension* (1904):

> If we pass a [three-dimensional] spiral through the ['two-dimensional'] film the intersection will give a point moving in a circle shown by the dotted lines in the figure. [...] If now we suppose a consciousness connected with the film in such a way that the intersection of the spiral with the film gives rise to a conscious experience, we see that we shall have in the film a point moving in a circle, conscious of its motion, knowing nothing of that real spiral the record of the successive intersections of which by the film is the motion of the point.[48]

Here Hinton modified the dimensional analogy to suggest that the consciousness of a being confined to a two-dimensional, plane surface is the result of movement in the third dimension of space (Figure 4.1). Thus it is *movement* that engenders consciousness. Similarly, in his 1904 statement of radical empiricism, William James claimed that consciousness does not 'exist' in the literal sense: consciousness is not an entity, James argued, but rather a *process*.[49]

---

47 Hinton to W. James, 5 October 1892, 1–2.
48 See Fig. 4.1, Hinton, *The Fourth Dimension*, 25.
49 See W. James, *Writings, 1902–1910*.

FOUR-DIMENSIONAL CONSCIOUSNESS

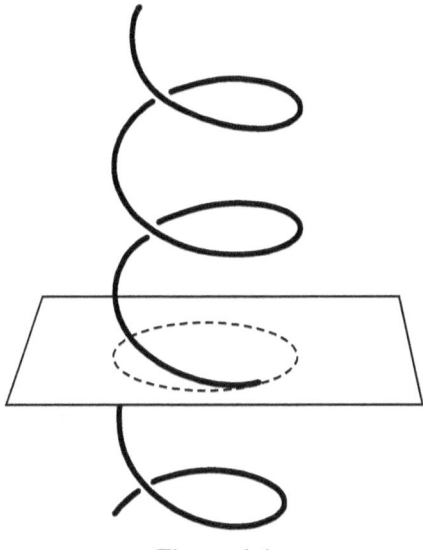

Figure 4.1

To complete the above dimensional analogy of the spiral and the film, Hinton continued:

> Passing to four dimensions and our space, we can conceive that all things and movements in our world are the reading off of a permanent reality by a space of consciousness. Each atom at every moment is not what it was, but a new part of that endless line which is itself. And all this system successively revealed in the time which is but the succession of consciousness, separate as it is in parts, in its entirety is one vast unity.[50]

The succession of states of consciousness, once conjoined, would reveal a unified, four-dimensional whole. It is fortuitous that Hinton described the tracing movement of the three-dimensional consciousness over the four-dimensional structure as 'reading', as seen below.

We can observe the connection between James's earlier psychological theory of consciousness as streaming and his later radical empiricism as limiting the definition of consciousness to a succession of experiences rather than as an entity in itself. In his 'Stream of Thought' chapter of the *Principles*, James highlighted the difficulty of using current epistemology (and the language by which it had been structured) to think about thinking. We might consider here the difficulty Hinton encountered in describing four-dimensional existence using the language of three-dimensional perception. After comparing the stream of consciousness to a bird's life of 'an alteration of flights and perchings', James wrote:

> Let us call the resting-places the 'substantive parts,' and the places of flight the 'transitive parts,' of the stream of thought. It then appears that the main end of our thinking is at all

---

50 Hinton, *The Fourth Dimension*, 26.

times the attainment of some other substantive part than the one from which we have just been dislodged. And we may say that the main use of the transitive parts is to lead us from one substantive conclusion to another.

Now it is very difficult, introspectively, to see the transitive parts for what they really are. If they are but flights to a conclusion, stopping them to look at them before the conclusion is reached is really annihilating them. [...] The attempt at introspective analysis in these cases is like seizing a spinning top to catch its motion, or trying to turn up the gas quickly enough to see how the darkness looks.[51]

Hence, we see the difficulty of language, particularly the English language, in attempting to express these transitive parts. In fact, James argued, 'large tracts of human speech are nothing but *signs of direction* in thought';[52] these are the 'if', the 'and', the 'but' and the 'by' of English, or, as Hinton called them, 'the waste products of thought'. While the substantive parts of thought (which correspond to the nouns, verbs, adverbs and adjectives of English grammar) are composed of 'sensorial images [that] are stable psychic facts', the function of the transitive parts, or 'psychic transitions', is 'to lead from one set of images to another'.[53] Returning to the stream-of-consciousness metaphor, James summed up his case by stating that 'the traditional psychology talks like one who should say a river consists of nothing but pailsful, spoonsful, quartpotsful, barrelsful, and other moulded forms of water. Even were the pails and the pots all actually standing in the stream, still between them the free water would continue to flow. It is just this free water of consciousness that psychologists resolutely overlook'. This is why James pushed for the 'reinstatement of the vague' to psychology, because

every definite image in the mind is steeped and dyed in the free water that flows around it. With it goes the sense of its relations, near and remote, the dying echo of whence it came to us, the dawning sense of whither it is to lead. The significance, the value, of the image is all in this halo or penumbra that surrounds and escorts it, – or rather that is fused into one with it and has become bone of its bone and flesh of its flesh; leaving it, it is true, an image of the same *thing* it was before, but making it an image of that thing newly taken and freshly understood.[54]

Just as it is impossible to step into the same river twice, it is impossible to think the same image twice. We can think an image of the same *object,* James argued, but each new thought is altered by the ones that have preceded it, making each encounter with the same object one of increased understanding. Here we see another anticipation of James's later pragmatist methodology of ambulatory relations. Consciousness thus works dynamically and cumulatively, by constant supplementation. This model of consciousness suited Hinton's hyperspace philosophy; as we saw in his early attempts to imagine the fourth dimension, the only way to construct this image was through a process of

---

51 W. James, *The Principles of Psychology*, 1: 243–44.
52 Ibid., 252–53.
53 Ibid., 253.
54 Ibid., 255, original emphasis.

continual supplementation, of returning to the thought of the fourth dimension with fresh experience.

The 'halo or penumbra' James described as surrounding each concrete image of consciousness is what Wolfgang Iser later identified in his description of the act of reading as 'provid[ing] the continual overlapping and interweaving of the views presented by each of the chapters', or sentences or words of a text.[55] Significantly, James depicted this halo or penumbra as a kind of aesthetic response; it was 'that shadowy scheme of the "form" of an opera, a play, or book, which remains in our mind and on which we pass judgment when the actual thing is done'.[56] Hinton's description of the creation of consciousness by a process of the three-dimensional brain 'reading' movements in the fourth dimension thus takes on another shade of meaning here. It is our aesthetic sense that allows us to approach the fourth dimension: if 'the life of the real [four-dimensional] being is read off again and again in successive waves of consciousness', then it is by somehow simultaneously holding the image of the entire 'sea' of consciousness in the attention that we achieve four-dimensional consciousness.[57]

While, according to James, 'We all of us have this permanent consciousness of whither our thought is going', 'this field of view of consciousness varies very much in extent'.[58] The scope depends on attentiveness, James argued, and those who possess what we call genius are naturally able to extend their field of consciousness the widest: 'Great thinkers have vast premonitory glimpses of schemes of relation between terms, which hardly even as verbal images enter the mind, so rapid is the whole process'.[59] James cited Mozart's statement of his own experience of this wide vista of consciousness, which reads like Ruskin's description of the imaginative artist *par excellence*:

> Mozart describes thus his manner of composing: First bits and crumbs of the piece come and gradually join together in his mind; then the soul getting warmed to the work, the thing grows more and more, 'and I spread it out broader and clearer, and at last it gets almost finished in my head, even when it is a long piece, so that I can see the whole of it at a single glance in my mind, as if it were a beautiful painting or a handsome human being; it which way I do not hear it in my imagination at all as a succession – the way it must come later – but all at once, as it were. [...] The best of all is the *hearing of it all at once*'.[60]

Hinton believed this variety of heightened aesthetic experience was accessible to all who possessed a four-dimensional consciousness. Mozart's simultaneous view of his entire composition is the same sort of suprahistorical perspective experienced by the narrator at the end of *An Unfinished Communication*. Hinton used the visual language of film here because the images of recurrence enter the narrator's mind with such rapidity that they

---

55 Iser, 'Indeterminacy', 39.
56 W. James, *The Principles of Psychology*, 1: 255.
57 Hinton, *The Fourth Dimension*, 27.
58 W. James, *The Principles of Psychology*, 1: 255–56.
59 Ibid., 255.
60 Ibid., 255, original emphasis.

are 'hardly even verbal'. In both cases, this wide consciousness allows for almost superhuman acts of creativity.

Unfortunately, according to James, such extraordinary feats of attentive consciousness seem to be limited to a very select number of people:

> Geniuses are commonly believed to excel other men in their power of sustained attention. In most of them, it is to be feared, the so-called 'power' is of the passive sort. Their ideas coruscate, every subject branches infinitely before their fertile minds, and so for hours they may be rapt. *But it is their genius making them attentive, not their attention making geniuses of them.*[61]

Even when Hinton's narrator manages to achieve this kind of hyper-conscious state of attentiveness in *An Unfinished Communication*, he does so passively after his accidental drowning. Hinton's decision to experiment with out-of-body experience in *An Unfinished Communication* may have been inspired by Schiller's suggestion that

> if the body is a mechanism for inhibiting consciousness, [...] it will be necessary also to invert our ordinary ideas on the subject of memory. It will be during life that we drink the bitter cup of Lethe [...]. And this will serve to explain [...] the extraordinary memories of the drowning and the dying generally.[62]

Schiller, pragmatist philosopher and (like Hinton) graduate of Balliol College, was a mutual acquaintance of William James and Hinton, and James cited this same passage in his discussion of the 'transmission theory' of mind, discussed in more detail below. As a transcendental *materialist*, however, Hinton was more interested in achieving this higher state of awareness while alive, and he believed the fourth dimension was the means of achieving this. The attainment of four-dimensional consciousness was not limited to geniuses only; Hinton's entire project was founded on the assumption that anyone who would undergo the arduous process of training the imagination could access the fourth dimension. What Hinton called for was a heroic act of will.

William James and the Heroic Will-to-Attention

As has been well-documented by his biographers and critics, a crucial turning point in William James's life occurred when he began to believe in the possibility of free will. In the spring of 1870, after a long period of battling depression and suicidal thoughts – what he later described as suffering as a 'sick soul' – he wrote his famous diary entry:[63]

> I think that yesterday was a crisis in my life. I finished the first part of Renouvier's second 'Essais' and see no reason why his definition of Free Will – 'the sustaining of a thought *because*

---

61 Ibid., 423, original emphasis.
62 Schiller, *The Riddles*, 296. W. James cites the same passage with approval in 'Human Immortality'.
63 I refer to James's discussion of his own depression, under the guise of a 'French correspondent' in 'The Sick Soul', in *Writings, 1902–1910*, particularly 149–51.

I choose to when I might have other thoughts' – need be the definition of an illusion. At any rate, I will assume for the present – until next year – that it is no illusion. My first act of free will shall be to believe in free will. [...] After the first of January, my callow skin being somewhat fledged, I may perhaps return to metaphysical study and skepticism without danger to my powers of action. For the present [...] I will go a step further with my will, not only act with it, but believe as well; believe in my individual reality and creative power.⁶⁴

James adopted the notion of free will as a useful fiction that allowed him to develop a sense of agency. By simply conceiving of free will, he created the possibility for its existence, and by acting as if free will exists, he confirmed its existence. By creating this useful fiction, James could become an author/creator in his own right. We see a similar crisis and resolution in Hinton's 'Casting Out the Self'. The fourth dimension, and particularly the prospect that we are four-dimensional agents, served as a useful fiction for Hinton, one that 'defend[ed] against a paranoid cognition of being psychically manipulated by another source of control', a sort of Cartesian deceptive demon.⁶⁵ Perhaps most disturbing for those thinkers who were just grasping toward the sense of something uncannily other within the human psyche – what Freud came to term the unconscious – was the possibility that this other source of control might come not from somewhere 'out there' beyond the consciousness, but from *within*. Hence, the need to bring the 'out there' of the spatial fourth dimension inside by making the self four-dimensional, as Hinton began to do in a self-conscious manner after his first series of *Scientific Romances*.

James's chapter on the will is one of the longest in his *Principles of Psychology*, and here again we see his struggle with the problem of free will. He decided that the question 'is insoluble on strictly psychological grounds', and if 'craving the sense of either peace or power' one feels the need to settle the matter, one must do so individually on a personal level. James stated that his own belief in free will was based on ethical rather than scientific considerations and thus was not appropriate to the context of the *Principles*. However, he continued, 'the logic of the question', was worth commenting upon.⁶⁶ James outlined what he called a '*fatalistic argument* for determinism': 'All is fate [...]. It is hopeless to resist the drift, vain to look for any new force coming in; and less, perhaps than anywhere else under the sun is there anything really mine in the decisions which I make'.⁶⁷ This argument deconstructs itself, James explained, because

there runs throughout it the sense of a force which might make things otherwise from one moment to another, if it were only strong enough to breast the tide. A person who feels the *impotence* of free effort in this way has the acutest notion of what is meant by it, and of its possible independent power. How else could he be so conscious of its absence and of that of its effects?⁶⁸

---

64 Quoted in Matthiessen, *The James Family*, 342, original emphasis.
65 Clarke, *Energy Forms*, 185.
66 W. James, *The Principles of Psychology*, 2: 572 and 573.
67 Ibid., 2: 574.
68 Ibid., original emphasis.

James argued that the sense of something lost or missing would be inconceivable to the true determinist; one would have to be at least aware of the possibility of free will in order to notice its absence. Therefore, it is 'not the *impotence* but the *unthinkability* of free-will' that true determinism affirms. There is a neat logic to James's argument here: we can hear echoes of Hinton's challenge to Kant's 'apodictic certainty' that space is limited to three dimensions in James's subversion of a fatalistic argument that 'strongly imagines the very possibility which determinism denies'.[69]

What is interesting in this discussion of determinism is James's invocation of the fourth dimension as the space in which the useful fiction of free will might be created. Determinism, James continued,

> admits something phenomenal *called* free effort, which seems to breast the tide, but it claims this is a *portion of the tide*. The variations of the effort cannot be independent, it says; they cannot originate *ex nihilo*, or come from a fourth dimension; they are mathematically fixed functions of the ideas themselves, which are the tide. Fatalism, which conceives of effort clearly enough as an independent variable that *might* come from a fourth dimension, if it *would* come but that it *does not* come, is a very dubious ally for determinism.[70]

Here the fourth dimension serves a rhetorical purpose; it is interchangeable with 'out of nowhere', described as a space that is as yet unlocated. However, it also represents a fiction engendered by a deep personal need to believe in free will, a need which James identified in himself, and likely recognized in Hinton as well.

In fact, the only time James cited Hinton in his published writings was in his essay, 'The Will to Believe', in which, among other things, James argued that an individual has a right – if no objective evidence exists either way – to act on the personal desire to believe one hypothesis over another. Elsewhere in 'The Will to Believe' James defined two ways of developing belief in any particular hypothesis: empiricist and absolutist. While the empiricist method reigns in scientific epistemology, in fact even in science 'the greatest empiricists among us are only empiricists on reflection: when left to their instincts they dogmatize like infallible popes'. Armed with the knowledge that 'we are all absolutists by instinct', and that all so-called empirical evidence is thus 'tainted' by our absolutist nature, what are we to do?

> Objective evidence and certitude are doubtless very fine ideals to play with, but where on this moonlit and dream-visited planet are they found? I am, therefore, myself a complete empiricist so far as my theory of human knowledge goes. I live, to be sure, by the practical faith that we must go on experiencing and thinking over our experience, for only thus can our opinions grow more true; but hold any of one of them – I absolutely do not care which – as if it never could be re-interpretable or corrigible, I believe to be a tremendously mistaken attitude.[71]

---

69 Ibid., original emphasis.
70 Ibid., original emphasis.
71 W. James, *Writings, 1878–1899*, 466.

Thus James's choice to undertake his ambulatory methodology was a conscious and ethical decision to continually attend to fresh evidence and revalue given truths in light of this evidence. Here James cited Hinton as one of the 'striking instances in point' of those who are willing to undertake this approach: 'the transcending of the axioms of geometry, not in play but in earnest, by certain of our contemporaries (as Zöllner and Charles H. Hinton)' served as an example of how 'we find no proposition ever regarded by anyone as evidently certain that has not either been called a falsehood, or at least had its truth sincerely questioned'.[72]

To remain continually attentive requires an act of will, according to James. As Levin observes, for James 'the will turns out to be little more than what James calls "effort of attention"'.[73] Those who are able to attend to a wide horizon of experience 'by grace of genius' are probably inhibited from 'acquiring habits of voluntary attention'. Thus, 'moderate intellectual endowments are the soil in which we may best expect, here as elsewhere, the virtues of the will, strictly so called, to thrive'.[74] The virtue of the will, according to James, 'seems to belong to an altogether different realm, as if it were the substantive thing which we *are*. [...] He who can make none is but a shadow; he who can make much is a hero'.[75] As we see in Chapter Six, Henry James endowed his 'rounded' centres of consciousness with this heroic will-to-attention, which differentiates them from the other, comparatively 'flat' characters in his fictions. As for William James, it was clear that the will-to-attention had moral implications: '*To sustain a representation, to think*, is, in short, the only moral act'.[76] The hero is one who, when faced with a difficult idea or experience will say, '*Yes, I will even have it so!*';[77] this was the closest James came to defining, according to Levin, 'our deepest conviction of selfhood [that] is thus a function of this effort of attention'.[78]

Hinton required a heroic will-to-attention in accessing the fourth dimension; this is 'a world which must be apprehended laboriously, patiently, through the material things of it, the shapes, the movements, the figures of it'.[79] In Hinton's later writings, this world was to be found not by looking somewhere 'out there' in or beyond the material world, but rather by turning inward to examine the 'aims and motives' of the self. This four-dimensional will is expressed 'in minute, almost invisible changes' within the individual.[80] Though Hinton turned the search inward, he also continued to develop new cube exercises. No longer interested in 'casting out the self' (as he was in the first series), in *The Fourth Dimension* he provided colour plates to guide the reader in painting the faces and

---

72 Ibid., 467. Prior to leaving England, Hinton had gone by his middle name, Howard. In the United States, he switched to Charles. This is perhaps another indicator of his attempt to avoid being associated with his bigamy scandal in England.
73 Levin, *The Poetics of Transition*, 61.
74 W. James, *The Principles of Psychology*, 1: 424.
75 Ibid., 2: 578, original emphasis.
76 Ibid., 2: 566, original emphasis.
77 Ibid., 2: 578, original emphasis.
78 Levin, *The Poetics of Transition*, 62.
79 Hinton, *The Fourth Dimension*, 3.
80 Hinton, *An Unfinished Communication*, 175.

terminal points of a series of cubes, which were then to be moved through various configurations in an attempt to represent the appearance of a four-dimensional tesseract as it passes through three-dimensional space. Hinton also attempted to construct a 'language of space' in which a monosyllabic utterance was assigned to each portion of each cube, in place of colour.[81] To successfully complete these exercises would indeed require a heroic act of attention,[82] and though he admitted a certain amount of artifice in their abstract nature, in the beginning:

> All I can do is to present to myself the sequences of solids, which would mean the presentation to me under my [three-dimensional] conditions of a four-dimensional object. All I can do is to visualise and tactualise different series of solids which are alternative sets of sectional views of a four-dimensional shape.[83]

By training the consciousness to hold successive images of the cube's movement in attention, perhaps one could achieve something that surpasses the Ruskinian act of imagination. If one achieved this through the cube exercises then, according to Hinton, they would discover that the soul 'has a four-dimensional experience'; for Hinton the soul was not something spiritual but, rather, 'corporeal and real, but with higher faculties than we manifest in our bodily actions'.[84] Trying to arrive at a similar transcendental materialist explanation of heightened experience, William James proposed a 'transmissive' theory of the brain, which he illustrated using a metaphor borrowed from another thinker who was interested in the fourth dimension, Gustav Theodor Fechner.

Hinton, James and Fechner's 'Mother-Sea' of Consciousness

It is striking that both James and Hinton used the metaphorics of fluidity in their discussions of consciousness. While undoubtedly this is indicative of the larger cultural moment of transition at the turn of the twentieth century, there is also evidence that both Hinton and James were influenced by an earlier, nineteenth-century, metaphor of consciousness, what James describes as the 'mother-sea' metaphor of Gustav Fechner. Fechner, as noted in Chapter One, was one of the first writers to use the dimensional analogy to express the concept of a spatial fourth dimension. In addition to this early contribution to hyperspace philosophy, Fechner is an important figure in the history of the development of psychology as a science. His theory of 'psychophysics', according to M. E. Marshall, 'provided experimental psychology with one its first and most lasting methods of quantitative description'; by demonstrating that the activities of the brain

---

81 See Hinton, *A Language of Space*. This text was originally published as a pamphlet and subsequently reprinted as an appendix to the second edition of *The Fourth Dimension*.
82 In 'The Education of the Imagination', in *Scientific Romances*, Hinton recalls his attempts to teach school boys 'cubical chess', where the flat chessboard is extended to three dimensions in the mind and the players must remember the positions of all pieces (18).
83 Hinton, 'A Language of Space', *The Fourth Dimension*, 255.
84 Ibid., 254 and 255.

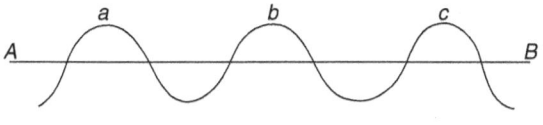

Figure 4.2

could be measured mathematically, Fechner laid the foundations for psychology as a branch of scientific – as opposed to strictly philosophical – experimentation and verification.[85] Like Charles Lutwidge Dodgson, Fechner published both professional, 'academic' texts as well as philosophical, romantic texts – he used the pseudonym 'Dr. Mises' for the latter variety. However, in the final years of his life, the interests of Fechner and Mises converged, and Fechner published in his own name a number of writings in mysticism.

William James, as Marshall demonstrates, was dismissive of Fechner and his theories during the years he was writing the *Principles*. 'You know I always thought his psychophysics as moonshiny as any of his other writings', James wrote to G. Stanley Hall in 1880.[86] However, during the 1890s, James began to read and appreciate Fechner for his philosophical and mystical ideas, as well as for the highly figurative language he used to express them.[87] It is Fechner's mother-sea metaphor in particular that gained James's later appreciation. This metaphor first appeared in the second volume of Fechner's *Elemente der Psychophysik* (1860), which has never been fully translated into English. James quoted from this text at length in his 1897 'Human Immortality' lecture, offering his own translation of the relevant passages. In the passage James cited, Fechner described his theory of the threshold of consciousness by using the image of a wave. James variously referred to this image as the 'wave scheme' and the 'mother-sea' of consciousness.

Fechner offered a diagram of a sinusoidal waveform criss-crossing over a horizontal line (Figure 4.2), explaining:[88]

> So far now as we symbolize any system of psycho-physical activity, to which a generally unified or principal consciousness corresponds, by the image of a total wave rising with its crest above a certain 'threshold,' we have a means of schematizing in a single diagram the physical solidarity of all these psycho-physical systems throughout Nature, together with their psycho-physical discontinuity.[...] In each wave the part that rises above the threshold is an integrated thing, and is connected with a single consciousness. Whatever lies below the threshold, being unconscious, separates the conscious crests, although it is still the means of physical connection.[89]

---

85 Marshall, 'William James', 304.
86 Cited in Perry, *The Thought and Character*, 1: 19; and. Marshall, 'William James', 304.
87 Perry, *The Thought and Character*, 2: 208.
88 See Fig. 4.2, Fechner, *Element der Psychophysik*, 2: 529; reproduced in W. James, *Writings, 1878–1899*, 1116.
89 Fechner, *Elemente der Psychophysik*, 2: 526–47, translated and abridged by W. James in *Writings, 1878–1899*, 1113–17, 1116, n. 6. Marshall observes that this is 'the very passage from the *Psychophysics* to which [William] had rather scathingly referred in the *Principles* apropos of Fechner's defence of mind-stuff' (307). It is not my intention here to offer a point-by-point

In Fechner's image, the crest of each wave represents an individual moment of human consciousness that is connected, below the surface, to all other waves through the larger body from which they are shaped. Using analogy, James expanded Fechner's metaphor from the individual conjunctive consciousness to a kind of pre-Jungian collective unconscious. James's interpretation of Fechner here is best given in his own words, which directly succeed his lengthy quotation of Fechner:

> One sees how easily on Fechner's wave-scheme a world-soul may be expressed. All psychophysical activity being continuous 'below the threshold,' the consciousness might also become continuous if the threshold sank low enough to uncover all the waves. The threshold throughout nature in general is, however, very high, so the consciousness that gets over it is of the discontinuous form.[90]

This citation of Fechner occurs within the context of James's explanation of the 'transmission theory' of mind–brain relations. According to James, in this theory, the brain has a 'permissive' or 'transmissive' function. Rather than assuming, as does the 'production theory' of materialism, that the brain *produces* consciousness on site as a kettle produces steam, supporters of the transmission theory argue that it is just as possible that the brain acts as a 'receiver' for consciousness which is actually produced and transmitted from somewhere beyond the physical brain.

James used Fechner's mother-sea metaphor to explain transmission theory. Lowering the threshold of attentive consciousness allows the individual to perceive its own continuity with the 'world-soul' or the unified source of consciousness from which other human brains also channel their individualized consciousness; by lowering this threshold, the shared foundation of each individual wave is allowed to surface. For James, the transmission theory had the advantage of allowing the existence of individual experiences that fall outside the ken of the production theory, such as intense religious experiences or the 'supernatural' experiences explored by psychical researchers: 'We need only to suppose the continuity of our consciousness with a mother-sea, to allow for exceptional waves occasionally pouring over the dam. Of course the causes of these odd lowerings of the brain's threshold still remain a mystery on any terms'.[91] Unlike James, Hinton thought he had found a way of intentionally lowering the threshold; he believed that, by undertaking his cube exercises, readers could train themselves to access the mother-sea, which can be read as the fourth dimension.

As has been well documented, William James was deeply involved with the American and British branches of the Society for Psychical Research, serving as president of the British SPR from 1893 to 1895. He personally studied several cases of alleged supernatural occurrence, and although he did not accept his subjects' descriptions of these experiences at face value, he found these personal encounters beyond natural 'reality'

---

examination of the contradictions between William's overall approach to psychology and Fechner's; however, a lucid discussion of these differences can be found in Marshall.
90 W. James, *Writings, 1878–1899*, 1117.
91 Ibid., 1118–19.

useful in studying consciousness. Like the student in 'The Persian King', who combined the folklore of the rural valley-dwellers with the scientific knowledge of the metropolis, James argued that one who acknowledged and examined the facts (i.e., case-studies of supernatural experiences) gathered by psychical researchers, 'while reflecting upon them in academic-scientific ways, will be in the best possible position to help philosophy'.[92] Thus, it is not surprising that James was drawn to Fechner's mother-sea metaphor; while remaining within the realm of science, it provided a means (like Hinton's fourth dimension) of conceptualizing how human consciousness might be able to access something that transcended the individual mind. It is in this light that we can understand why Hinton's first romance, 'What Is the Fourth Dimension?' was subtitled 'Ghosts Explained'.

The fact that Hinton's wife, Mary, was inspired to write a poem based on Fechner's *Zend-Avesta* (1851) strongly suggests that Hinton was aware of Fechner and his work as well, and it seems likely that Hinton had Fechner's mother-sea metaphor in mind when he was writing *An Unfinished Communication*.[93] Here, in the initial conversation between the Unlearner and the narrator in which the Unlearner distinguishes between unlearning and forgetting, the Unlearner employs a similar metaphor. 'Look at the sea', he instructs the narrator as they stand together on coastal path:

> From here we can see a multitude small waves; if we were on a high eminence we should see the larger ocean billows on whose surface merely these small disturbances are. From a still greater height we should see the great wave of the tide, whose great sweep might mean life or death to a swimmer, buffeting the little waves.[94]

At this point in *An Unfinished Communication* the narrator desires oblivion through forgetting, rather than *amor fati* through unlearning. The Unlearner argues that the self-transcendence the narrator seeks is accomplished not through self-annihilation, but through *living*, a process he defines: 'Life is where man takes up the work of nature and forms a net-work of close personal knowledge, linking each to each, preparing that body in which the soul of man lives'.[95]

The Unlearner offers a pragmatist definition of the process of living as conjunctive, which is in line with William James's ambulatory methodology: 'I live, to be sure, by the practical faith that we must go on experiencing and thinking over our experience, for only thus can our opinions grow more true',[96] or to put it more simply, as James does ten years later, 'when [...] asked what truth *means*, we reply by telling only how it

---

92 W. James, *Writings, 1878–1899*, 682.
93 See M. B. Hinton, 'After Death (Found on Passage in the Zend Avesta)', in *Other Notes* 40. The reference here is to Fechner's *Zend-Avesta* (1851), which has never been completely translated into English. The text was reissued in German in 1901, and fragments of it have been translated into English after this date, for example in appendices to Fechner's *The Little Book of Life after Death*.
94 Hinton, *An Unfinished Communication, Scientific Romances*, 119.
95 Ibid., 120.
96 W. James, *Writings, 1878–1899*, 466.

is *arrived-at*.⁹⁷ Both James and the Unlearner could only express how truth is arrived at rather than what it means, because truth is individually and continually created through the process of mental ambulation. Hinton dramatized this methodology in the eternal recurrence sequence of *An Unfinished Communication*, in which, after drowning in the sea, the narrator experiences and re-experiences all possible permutations of his life: 'In my higher consciousness I see wherefrom, whereunto it [his four-dimensional life] moulds itself. I see how each little thing is different, and how in just this, that now I have lived, I grasp the realized results of ages of the higher transverse growth'.⁹⁸ It is in grasping the conjunctive relations – the 'wherefrom' and 'whereunto' – that Hinton's narrator accesses the fourth dimension.

Returning to the Unlearner's sea metaphor quoted above, in addition to foreshadowing the narrator's death by drowning, it directly echoes Fechner's mother-sea metaphor, as translated by James:

> We may represent such a long period as that of the slowly fluctuating condition of our general wakefulness and the general direction of our attention as a wave that slowly changes the place of its summit. If we call this the *under-wave*, then the movements of the shorter period, on which the more special conscious states depend, can be symbolized by wavelets superposed upon the under-wave, and we can call these *over-waves*. They will cause all sorts of modifications of the under-wave's surface, and the total wave will be the resultant of both sets of waves. [...]
>
> In each wave the part that rises above the threshold is an integrated thing, and is connected with a single consciousness.⁹⁹

The over-waves, or the waves that appear above the threshold of consciousness, are states of consciousness within each individual that may appear to be separate, but are actually connected beneath the threshold. If we transpose this metaphor for the individual subject onto humanity as a whole (as James and Hinton did), then the over-waves represent individual minds, which are in turn connected beneath the threshold.¹⁰⁰

For both James and Hinton, Fechner's mother-sea metaphor played a key role in expressing their sense that there is something 'out there' beyond apparent physical reality but accessible through the human psyche. In his discussion of Fechner, William James constructed his own metaphor that anticipated his brother Henry's famous 'house of fiction': 'Our brains are colored lenses in the wall of nature, admitting light from the super-solar source, but at the same time tingeing and restricting it'.¹⁰¹ In the transmission theory, the brain works as a framing device; it is a lens that both focuses and distorts.

---

97 W. James, *Writings, 1902–1910*, 904, original emphasis.
98 Hinton, *An Unfinished Communication*, *Scientific Romances*, 176.
99 W. James, *Writings, 1878–1899*, 1116, n. 6, original emphasis.
100 An insightful initial exploration of Fechner's influence on Walter Pater has been made by Meisel in 'Psychoanalysis and Aestheticism'. Indeed, there is scope for examining the recurrence of his sea metaphor in high modernist fiction, particularly Virginia Woolf's *The Waves*. Unfortunately, this is beyond the focus of this book.
101 W. James, *Writings, 1878–1899*, 1112.

As Hinton wrote to James in 1892, he was working under the assumption that human beings are themselves 'indefinitely flat in the fourth dimension', meaning that humans do extend (however minutely) into the fourth dimension, but are unaware of it. He claimed this limitation is one of consciousness and this idea takes on new meaning when considered alongside James's discussion of the transmission theory. Hinton's theory that human beings are four-dimensional is in fact a correlative to transmission theory, and Fechner's mother-sea metaphor is a useful way of imagining the individual's experience of four-dimensional consciousness.

For both James and Hinton the mother-sea metaphor also served as a means of conceptualizing the possibility of a unified 'over' consciousness in which all human beings partake. While perhaps a more appropriate prefix would be 'under', especially in light of Fechner's image of the 'under-wave' in which separate consciousnesses find their unity (or indeed, Freud's later work on the *sub*conscious) James, like Hinton, continued to use the trope of *over*-ness, of height or transcendence. In 1909, James revisited Fechner's mother-sea metaphor, citing the case studies of Freud and other contemporary psychologists:

> I find in some of these abnormal or supernormal facts the strongest suggestions in favor of a *superior* co-consciousness being possible. I doubt whether we shall ever understand some of them without using the very letter of Fechner's conception of a great reservoir in which the memories of the earth's inhabitants are pooled and preserved, and from which, when the threshold lowers or the valve opens, information ordinarily shut out leaks into the mind of exceptional individuals among us.[102]

Thus, for James there is a much-wider realm of consciousness that humans may occasionally experience, but it is accessible only under extraordinary circumstances. Before Freud's theorization of the subconscious, for James this 'superior or co-consciousness' was 'another dimension of being'.[103] It was in his discussions of this other dimension of being that James most closely approximated Hinton's language: 'May not you and I be confluent in a higher consciousness and confluently active there, tho we know it not?', he asked in 1908.[104] Hinton sought a way of 'knowing' and experiencing this higher consciousness, which he conceived of as four-dimensional. Much early twentieth-century fiction is underpinned by this same belief in something 'more' within the human psyche, and for some writers and artists, 'the fourth dimension' continued to carry cultural currency as an effective shorthand, a means of reinstating 'the vague and inarticulate to its proper place in our mental life'.

---

102 W. James, *1902–1910*, 766, emphasis added.
103 Levin, *The Poetics of Transition*, 65.
104 W. James, *Writings, 1902–1910*, 762.

Chapter Five

H. G. WELLS'S FOUR-DIMENSIONAL LITERARY AESTHETIC

When exploring the influence of Hinton's ideas at the turn of the twentieth century, most scholars look to H. G. Wells; he called his own proto-science fictions 'scientific romances', after all. Wells's early writings demonstrate his interest in the theory of the fourth dimension, an interest that led to the development of what William J. Scheick has called Wells's 'splintering frame technique'.[1] While Scheick provides a useful framework for rethinking Wells's later writing, many examinations of Wells's early work are rooted in the reductive 'history of ideas' model of criticism. In this chapter, I traverse a selection of Wells's early texts, with an eye not only to Hinton's influence, but to the nascent modernist 'history of consciousness' in which both Hinton and Wells participated.[2] For both men, the theory of the fourth dimension appealed as a means of explaining and harnessing the conflicting and disruptive forces unleashed by scientific and technological developments of the nineteenth century, forces that were becoming a source of increasing anxiety at the turn of the century.

I focus primarily on Wells's work up to 1915, from his earliest scientific romances up to his controversial book, *Boon*, where he publicly attacked Henry James and his literary aesthetic. To begin, I examine his earliest uses of four-dimensional theory in *The Time Machine* (1894–1895) and *The Invisible Man* (1897). I focus particularly on *The Invisible Man*, a text that was clearly influenced by Hinton's *Stella* (1895). It was to Wells's advantage that he began composing *The Invisible Man* shortly after Wilhelm Conrad Röntgen discovered X-rays late in 1895.

Like Hinton, Wells turned to cutting-edge visual technologies in his attempt to represent four-dimensional space and subjectivity, and the unprecedented public enthusiasm for X-ray imaging provided Wells with a powerful aid in imagining the unimaginable. X-ray images in particular seemed to provide material evidence that there was a fourth dimension of space, a perspective from which three-dimensional objects were transparent. Tom Gibbons has observed that 'X-rays [...], and their contribution to the anti-materialistic millenarian synthesis, appear largely responsible for the continuous excitement about the Fourth Dimension among the general public and avant-garde painters alike' at the turn of the twentieth century.[3]

---

1 See Scheick, *The Splintering Frame*.
2 See Hocks, *Henry James*, 43.
3 Gibbons, 'Cubism', 140.

Like William James, Wells was intrigued by the sense of 'more', of something beyond the capabilities of unaided human perception. In 1904, he wrote of the need for 'scepticism of the instrument'; in question here is 'the Instrument of Human Thought', as expressed through scientific theories and accumulated cultural assumptions.[4] In his revision of this essay for inclusion as an appendix to his 1905 experimental fiction *A Modern Utopia* (the same text that prompted William James to write his first letter to Wells in praise of the work), Wells added the claim that '*the forceps of our minds are clumsy forceps, and crush the truth a little in taking hold of it*'.[5] Wells was intrigued by the concept of the fourth dimension; he believed that within this idea lay the means for fine-tuning the instrument of human consciousness to transcend itself. Many of Wells's experiments in fiction can be read as attempts to develop a four-dimensional literary aesthetic, one that would engender a higher level of consciousness in his readers.

Four-Dimensional Invention

Wells's first scientific romance, *The Time Machine: An Invention*, appeared as a serial publication in the *New Review* in 1894 and as a novella in 1895. Wells had experimented with the idea of time travel even earlier, in his 1888 short story, 'The Chronic Argonauts', published in the *Science Schools Journal* shortly after Wells was a student at the Normal School of Science at Kensington. In both texts Wells used Hinton's theory of a spatial fourth dimension, experienced as duration, as a jumping-off point for the narrative. In *The Time Machine*, after offering a modified version of the dimensional analogy of the relations between a line, a plane and a cube, Wells's nameless Time Traveller asks the guests at his dinner-party: 'Can an instantaneous cube exist?'. The answer is no, of course, because:

> 'Clearly' the Time Traveller proceeded, 'any real body must have extension in *four* directions: it must have Length, Breadth, Thickness, and – Duration. But through a natural infirmity of the flesh, which I will explain to you in a moment, we incline to overlook this fact. There are really four dimensions, three which we call the three planes of Space, and a fourth, Time. There is, however, a tendency to draw an unreal distinction between the former three dimensions and the latter, because it happens that our consciousness moves intermittently in one direction along the latter from the beginning to the end of our lives. [...] *There is no difference between Time and any of the three dimensions of our Space except that our consciousness moves along it.*'[6]

According to Wells's Time Traveller, we *call* the fourth dimension 'time', but in fact our separation of it from the three known dimensions of space is artificial; it is a limitation of embodiment. Like Hinton, the Time Traveller argues that the fourth dimension is experienced as duration, but in reality it is the movement of our consciousness through a space it cannot fully recognize and manipulate. And, like Hinton, Wells looked to the disorienting visual spectacles provided by innovations in transportation and moving-picture

---

4   Wells, 'Scepticism', in *Mind*, 389.
5   Wells, 'Scepticism', in *A Modern Utopia*, 256, original emphasis.
6   Wells, *The Definitive Time Machine*, 32, original emphasis.

technology to represent movement through the fourth dimension. Aboard his bicycle-like time machine, the Time Traveller compares four-dimensional movement to riding 'upon a switchback' (another word for roller-coaster, which was first patented in 1885), and the rapid passage of time all around him resembles the unnatural spectacle of stop-motion photography:

> As I put on pace, night followed day like the flapping of a black wing. […] I saw the sun the saw hopping swiftly across the sky, leaping it every minute, and every minute marking a day. […] The slowest snail that ever crawled dashed by too fast for me. The twinkling sensation of darkness and light was excessively painful to the eye. […] Presently, as I went on, still gaining velocity, the palpitation of night and day merged into one continuous greyness […] the jerking sun became a streak of fire; a brilliant arch, in space; the moon a fainter fluctuating band; and I could see nothing of the stars, save now and then a fainter circle flickering in the blue.[7]

Keith Williams observes, 'Commentators agree [that] what makes Wells's descriptions of the visual effects of time travel truly extraordinary is that he wrote them *before* he could possibly have seen a film, at least in the cinematograph's sense of the public screening'.[8] These commentators are correct; however, as I have shown, Wells was not the first to draw on filmic narrative to represent movement through the fourth dimension.

In his earlier romance of time travel, written before the publication of Hinton's recurrence scene in *An Unfinished Communication*, Wells did not attempt to depict the actual experience of time travel. He did, however, use Hinton's fourth dimension again in his explanation of the possibility of time travel. In 'The Chronic Argonauts' his time traveller, Dr Moses Nebogipfel, first uses the dimensional analogy to explain the fourth dimension and then continues:

> When we take up this new light of a fourth dimension and reexamine our physical science in its illumination […] we find ourselves no longer limited by hopeless restriction to a certain beat of time – to our own generation. Locomotion along lines of duration – chronic navigation comes within the range, first, of geometrical theory, and then of practical mechanics.[9]

Nebogipfel expresses Wells's nascent pragmatist methodology here; it is by revisiting and re-evaluating experience in light of a new understanding of the fourth dimension that great mechanical and intellectual advancements are made. This is, in fact, how Nebogipful has managed to manipulate his 'locomotion' in the fourth dimension, a monumental development which he positions alongside other paradigm-shifting discoveries. However, Nebogipfel, with this new power, is also threatening:

> There *was* a time when men could only move horizontally and in their appointed country. The clouds floated above them, unattainable things, mysterious chariots of those fearful gods who dwelt among the mountain summits. Speaking practically, men in those days were restricted

---

7 Ibid., 42.
8 K. Williams, *H. G. Wells, Modernity and the Movies*, 27, original emphasis.
9 Wells, *The Definitive Time Machine*, 150.

to motion in two dimensions; and even there circumambient ocean and hypoborean fear bound him in. But those times were to pass away. First, the keel of Jason cut its way between the Symplegades, and then in the fulness of time, Columbus dropped anchor in a bay of Atlantis. Then man burst his bidimensional limits, and invaded the third dimension, soaring with Montgolfier into the clouds, and sinking with a diving bell into the purple treasure-caves of the waters. And now another step, and the hidden past and unknown future are before us. We stand upon a mountain summit with the plains of the ages spread below.[10]

There is a note of megalomania in Nebogipfel's use of the trope of 'over-ness' to describe his position in relation to the rest of humanity, and the comparison of his access to the 'mountain summit' with the power of 'fearful gods'. Nebogipfel is an ambiguous character at best: he is sinister in physical appearance and, we learn at the end of the narrative, guilty of manslaughter at least, if not outright murder. Wells's later Time Traveller of *The Time Machine* is more positively portrayed as a gentleman scientific investigator; his desire to harness the fourth dimension is driven by curiosity rather than thirst for power. Wells did not return to the threatening implications of the powerful, four-dimensional subject until *The Invisible Man*, which I discuss later in this chapter.

There is further evidence that Wells had Hinton's hyperspace philosophy in mind in *The Time Machine*. Though he does not mention Hinton by name, his Time Traveller alludes to Hinton's cube exercises:

> You know how on a flat surface, which has only two dimensions, we can represent a figure of a three-dimensional solid, and similarly they think that by models of three dimensions they could represent one of four – if they could master the perspective of the thing.[11]

In his attempts to master this perspective, Hinton developed his 'Arabic method' aesthetic in the *Scientific Romances*, as well as his complex cube exercises. The Time Traveller's description of the 'models of three dimensions' here echoes Hinton's discussion of his models in 'Casting Out the Self' (1886):

> The whole block of cubes formed a kind of solid [three-dimensional] paper in which one could mentally put down any solid shape one wanted. And just as it is a great convenience to have a piece of paper in drawing figures one wants to think about, so it was a great convenience to have this solid paper.[12]

Wells also utilized Hinton's representation of the fourth dimension as a series of three-dimensional slices. His Time Traveller explains:

> Well, I do not mind telling you I have been at work upon this geometry of Four Dimensions for some time. Some of my results are curious. For instance, here is a portrait of a man at eight years old, another at fifteen, another at seventeen, another at twenty-three, and so on.

---

10 Ibid., 150, original emphasis.
11 Ibid., 32.
12 Hinton, 'Casting Out', *Scientific Romances*, 226.

All these are evidently sections, as it were, Three-Dimensional representations of his Four-Dimensioned being, which is a fixed and unalterable thing.[13]

Keith Williams argues that here Wells wrote 'as if imagining the course of a life as one vast chronophotograph' using terminology 'similar to Henri Bergson's contemporary theory of consciousness and memory'; in fact, Hinton's attempts to represent the fourth dimension seem a more plausible influence.[14] Wells owned a copy of Hinton's *Scientific Romances*, and he was a member of the Student Debating Society at the Normal School where Hinton's ideas were the main topic of at least one meeting.[15]

While Williams rightly observes that, in his early fictions, Wells 'strove to imagine forms and techniques of time-based representation that did not, as yet, exist in actual examples from often ingenious but relatively primitive cinema shows of the period', Wells, like Hinton, utilized new visual technologies in an attempt to represent something as yet unimaginable within literary narrative.[16] This ekphrastic striving is what marks the scientific romances of both Wells and Hinton as nascent modernist fictions: their affinity with experiments in early moving-picture technology is not a case of influence or anticipation, but rather of shared fascination with the possibilities of extending human perception. Film, like X-ray radiographs, could capture and make visible material realities previously inaccessible to the human eye. Wells and Hinton recognized the potential of these new technologies for expressing the fourth dimension.

The *Other* Invisible Protagonist

Williams argues that *The Invisible Man* is 'Wells's most sustained proto-cinematic exploration' of the late Victorian crisis of transgression of the boundaries between the seen and unseen.[17] While I find it useful to consider *The Invisible Man* within the context of contemporary developments in visual technology (especially given the spectacular nature of this novella's protagonist), I believe a more fruitful examination of this text will look to its relationship with Hinton's work and the 'four-dimensional' potential of X-ray imaging. In his exploration of the effect of invisibility on the individual subject, it is clear that Wells was responding to Hinton's *Stella* by engaging with questions of social, gendered

---

13 Wells, *The Definitive Time Machine*, 33.
14 K. Williams, *H. G. Wells, Modernity and the Movies*, 24. The theory Williams seems to have in mind here is expressed in Bergson's *Matter and Memory* (1896), which was not published until after *The Time Machine*. When Wells was writing *The Time Machine*, Bergson was just beginning his career, having only recently published his doctoral thesis, 'Time and Free Will' in 1889. This text was not available in English until 1910. There is no evidence that Wells was able to read French well enough to understand Bergson's argument.
15 See Hamilton-Gordon, 'The Fourth Dimension'. Hamilton-Gordon first presented this paper at the Student Debating Society, claiming the idea as his own. He later admitted to having read Hinton.
16 K. Williams, *H. G. Wells, Modernity and the Movies*, 7.
17 Ibid., 13.

identity; additionally, Wells extended Hinton's experiment in *Stella* to a consideration of the 'othering' effect of the fourth dimension.

As noted in the introduction, Wells presented the four-dimensional Angel of *The Wonderful Visit* (1895) as a Wildean aesthete. The implication here is that access to a four-dimensional consciousness somehow marks the subject as 'other' and, like Wilde, the Angel is denounced as a degenerate and socialist by the small-minded inhabitants of the village of Sidderton, where he was literally shot down to earth. Interestingly, Ernest Newman described Wilde's aesthetic 'immorality' as being shaped by his 'view from the fourth dimension'.[18] Indeed, there is perhaps something 'queering' about the fourth dimension, as Hinton acknowledged in 'A Plane World' (1886).

In this text from his first series of *Scientific Romances*, Hinton observed that 'in every man there is something of a woman, and in every woman there are some of the best qualities of a man. But in the [two-dimensional, 'plane'] world of which we speak there is no physical possibility for such interfusion'.[19] Hinton described a two-dimensional world of upright, right-angled, triangle-shaped inhabitants. The 'feet' of each being are located at the 'bottom', or horizontal, cathetus of the triangle. The 'face', or 'sensitive edge' is located on the vertical cathetus, and the 'hard edge' of the being is located along the hypotenuse: 'on the sensitive edge is the face and all the means of expression of feeling. The other [hypotenuse] edge is covered with a horny thickening of the skin, which at the sharp point becomes very dense and as hard as iron'. The 'female' triangles are mirror images of the 'male' triangles. Hinton even provided paper cut outs for the reader to utilize in imagining these beings: 'It must be remembered that the figures cannot leave the plane on which they are put. They must not be turned over' (Figure 5.1).[20]

The shape of the triangles determines their sex and their relations with one another, which are necessarily heterosexual:

> It is evident that the sharp point of one man is always running into another man's sensitive or soft edge. [...] It will be evident, on moving the figures about that no two men could naturally come face to face with each other. In this land no such thing as friendship or familiar intercourse between man and man is possible. The very name of it is ridiculous to them.[21]

However, Hinton reported, there is a 'curious history' in the 'annals of this race': a male triangle, Vir, and female triangle, Mulier, who had been 'living in a state of utmost perfect happiness' are disturbed one day when, 'owing to certain abstruse studies of the Mulier, she was suddenly, in all outward respects, turned irremediably into a man'. This bizarre transformation was the result, according to Hinton, of the scientific investigator Mulier somehow gaining access to the third dimension of space and physically flipping herself over while there. She eventually manages to reverse the effects of her experiment, but in the interim, she and her partner continue their relationship in a

---

18 See Newman, 'Oscar Wilde'.
19 Hinton, 'A Plane World', *Scientific Romances*, 140.
20 Ibid., 141–43. See Fig. 5.1, Ibid., 141.
21 Ibid., 145.

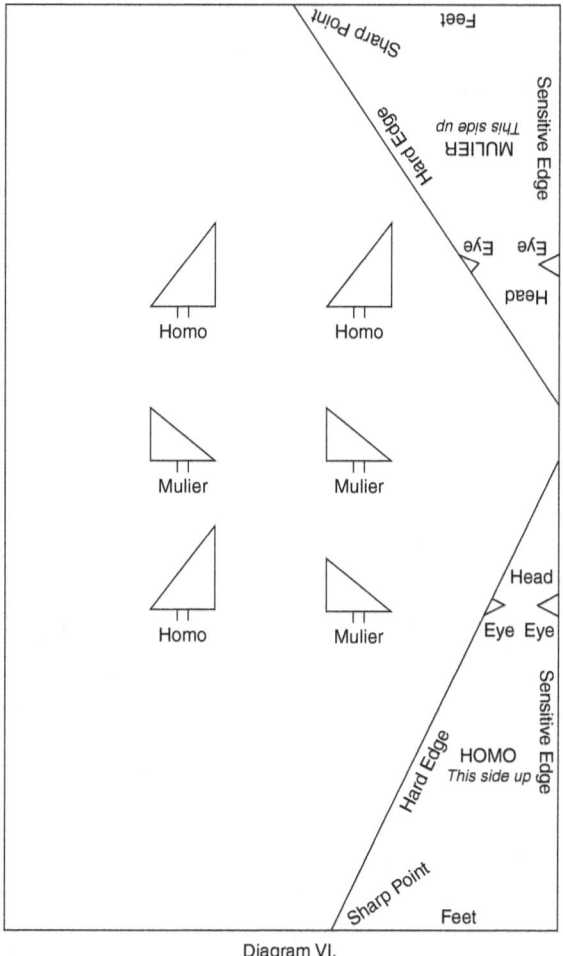

Diagram VI.

Figure 5.1

fashion that can only be read as crudely 'homosexual': 'Vir recognized her as the same true Mulier. But she occupied the same position with regard to him that any other man would. It was only by standing on his head that he could, with his sensitive edge, approach her sensitive edge'.[22]

Hinton made an odd choice here: why have a female scientist character at a time when female scientists were a rarity? This choice does not serve any particular plot function within the narrative. However, if we consider that 'A Plane World' was published shortly after the passage of the Criminal Law Amendment Act of 1885, which outlawed homosexual relations between men but not women, Hinton's choice becomes clear. Positioning 'Vir' as the scientific investigator who manages to flip himself over in the

---

22 Ibid., 146.

third dimension and thus returns as a physical woman who proceeds to continue 'her' relationship with Mulier as a 'sexual invert' would not have the same impact because contemporary Victorian law did not even acknowledge the significance or possibility of a relationship between two women. Just as his friend, Havelock Ellis, and the social progressives of fin-de-siècle London offered analyses of 'sexual inversion', 'A Plane World' appears to imply a 'scientific' explanation for homosexuality, one that mitigates moral condemnation in favour of tolerance or, at the very least, pity.[23] Ten years after the publication of 'A Plane World', Wells explored the possibility of a three-dimensional human 'flipping' in the fourth dimension in 'The Plattner Story' (1896). Although his protagonist returns permanently physically inverted (for example, his heart is now on the right side of his chest), unlike Hinton, Wells did not pursue the sexual implications of 'inversion' here. However, a year later in *The Invisible Man*, Wells did pick up on the 'queering' effect of male invisibility.

Wells's invisible man, Griffin, is a scientific investigator – like Hinton's Michael Graham – working in the 'border land' between scientific disciplines: as he explains to a colleague, he 'dropped medicine and took up physics' because 'light fascinated me'. During the course of his research, Griffin found 'a general principle of pigments and refraction, – a formula, a geometrical expression involving four dimensions'.[24] Graham rendered Stella transparent by lowering her 'coefficient of refraction [...] equal to one'.[25] Similarly, Griffin's invisibility was caused by 'lower[ing] the refractive index [...] to that of air'.[26] In *Stella*, Frank Cornish demonstrates how it might be possible to render a human being transparent when he conducts an initial experiment on a piece of flesh, and 'by immersing it in a heavy oil of the same coefficient of refraction [...], and keeping it under the air-pump for a long time, I permeated the minute passages; the result is a substance invisible in the oil, but which looks like a piece of glass out of it'.[27] Griffin offers a simpler version of this experiment in his explanation of invisibility to a colleague, Dr Kemp: 'Oil white paper, fill up the interstices between the particles with oil so that there is no longer refraction or reflection except at the surfaces, and it becomes as transparent as glass'.[28] Wells even used a similar argument to overcome an objection to the possibility of making a living human invisible. In *Stella*, after Cornish demonstrates the process with the oil and flesh, Churton notes that because of the blood circulating in live tissues, 'you could not treat a living person so'. Cornish explains that 'the blood owes its colour to salts of iron [...] all we have got to do is replace the iron by some element having colourless compounds'.[29] This is, admittedly, a scientifically dubious resolution to the problem, as is Wells's. Griffin anticipates a similar objection from Kemp by stating: 'You know the red colouring matter of the blood; it can be made

---

23 See, for example, Ellis and Symonds, *Sexual Inversion*.
24 Wells, *The Invisible Man*, 89.
25 Hinton, *Stella, Scientific Romances*, 34–35.
26 Wells, *The Invisible Man*, 89.
27 Hinton, *Stella, Scientific Romances*, 55–56.
28 Wells, *The Invisible Man*, 91.
29 Hinton, *Stella, Scientific Romances*, 56.

white – colourless – and remain with all the functions it has now!'.[30] Thus Wells's 'science' mimicked Hinton's.

However, Wells's interest in *Stella* went beyond the 'scientific patter' that provided an explanation for human invisibility.[31] In Hinton's story, Graham decided to make Stella invisible in order to break down the limitations of her three-dimensional subjectivity that had been formed by her gendering as feminine and, therefore, an aesthetic object. The idea that invisibility is a gender-specific remedy for the limitations of three-dimensional consciousness is explicit throughout *Stella* Invisibility would not be an appropriate treatment for men because, according to Churton, unlike female vanity, male limitations are expressed as more vigorous urges such as 'grabbing things and fighting'.[32] Applied to *The Invisible Man*, this becomes an apt description of much of Griffin's behaviour.

If self-effacement makes Stella a docile and altruistic individual, it turns Griffin – a narcissistic loner – into a sociopath. In Hinton's narrative, Stella's lack of self-regard makes her an emblem of Graham's socialist philosophy of 'being for others'. Conversely, Griffin's invisibility further isolates and frustrates him, pushing him to become the most extreme sort of individualist.[33] Dr Kemp's description of the invisible Griffin inverts the language of Graham's philosophy of 'being for others':

> He is mad, [...] inhuman. He is pure selfishness. He thinks of nothing but his own advantage [...]. I have listened to such a story this morning of brutal self-seeking! [...] The man's become inhuman [...]. He has cut himself off from his own kind.[34]

Griffin's reaction to his invisibility develops and pushes a brief episode from *Stella* to its logical conclusion. In this incident Stella tries to convince Churton to drink the invisibility drug. While they are abroad, Stella suggests that Churton become transparent because then 'we shall be like one another. Won't that be nice?'.[35] Churton reacts with outrage and paranoia:

> Now, during my walks about [Hong Kong], I had occasionally seen the faces of some men I had known in London, who had come out to posts in the Cingalese civil service. They had not recognised me, and this I had put down to their not expecting to see me there. [...] But now it flashed upon me that I might have been getting transparent all this while – that perhaps my face was a sort of mist. 'Good Heavens, Stella!' I exclaimed, 'you haven't been giving me any of that drink before, have you?'[36]

---

30 Wells, *The Invisible Man*, 92.
31 This is Wells's term; see his introduction to *The Scientific Romances of H. G. Wells*, vii–x.
32 Hinton, *Stella, Scientific Romances*, 83.
33 See Wells's use of the term *individualism* to describe 'the cult', which is underpinned by the 'essential fallacy' of the rejection of the principle that 'liberty is a compromise between our own freedom of will and the wills of those with whom we come into contact' (*A Modern Utopia*, 29).
34 Wells, *The Invisible Man*, 127.
35 Hinton, *Stella, Scientific Romances*, 82.
36 Ibid., 81.

Stella assures him that she has not, and after demanding she 'take that rubbish away', he regains his composure.[37] However, Churton's momentary uncertainty of his own physical appearance and its impression on other – specifically male – onlookers, is emasculating. He does not want to be 'like' Stella (or any woman). By rendering a male subject invisible in *The Invisible Man*, Wells created a protagonist for whom the signifiers of physical appearance become, by necessity, an obsession. The ambiguity surrounding Griffin exists not only in relation to his fantastical (lack of) appearance; he also behaves in gendered extremes. Griffin, like his namesake, the gryphon, is composed of two opposing parts, and it is his inability to reconcile these differences that ultimately leads to his destruction. While Stella is assimilated into the role of wife and mother, Griffin is killed in order to allow the return to the status quo.

For Griffin, to become invisible is to surrender 'the hegemonic heterosexual male body' that Annie Potts identifies as 'self-contained [...] with its exteriorised sexuality personified in the penis-self'. This body is conceptualized in opposition to

> the incoherence and interchangeability of the feminized body, whose orifices represent thresholds, margins of error, sites of weakness where outside may infiltrate inside, and vice versa. This 'male model' of sexuality is 'out there' [...]: the privileging of vision over other senses reifies the penis as an external organ.[38]

His body no longer visibly 'out there', Griffin is repeatedly othered, or gendered as feminine. Aside from his new obsession with his physical appearance, Griffin explains how his invisible status makes it nearly impossible for him to travel unaccompanied in London: 'Every crossing was a danger, every passenger a thing to watch alertly. One man as I was about to pass him at the top of Bedford Street, turned upon me abruptly and came into me, sending me in the road and under the wheel of a passing hansom'.[39] As an outsider to physical norms and cultural expectations, Griffin is denied agency, similar to the kind of effacement experienced by the unescorted, highly-sexualized 'fallen' Victorian woman or, indeed, women in general under much of nineteenth-century law.

Perhaps most striking is Griffin's newly problematic relationship with food and the act of eating. Even when disguised he is unable to eat in public, lest he reveal his invisible mouth, and when undressed he must fast, 'for to eat, to fill myself with unassimilated matter, would be to become grotesquely visible again'.[40] Like his female counterpart, 'eating is [...] something too personal to survive the public scrutiny of the dinner table'.[41] Helena Michie has highlighted the obsessive linking of food with female sexuality in nineteenth-century fiction, observing that Victorian heroines are rarely depicted in the act of eating. After becoming invisible, Griffin too must develop coping strategies, 'elaborate rituals' around the act of eating: even when dining privately with Dr Kemp, his sole

---

37 Ibid., 82.
38 Potts, *The Science/Fiction of Sex*, 203.
39 Wells, *The Invisible Man*, 115.
40 Ibid., 114.
41 Michie, *The Flesh*, 20.

confidant, Griffin demands a dressing gown, explaining that 'I always like to get something about me before I eat [...]. Queer fancy!'.⁴²

Dressing and 'painting' – also traditionally feminine activities – become obsessions for the invisible Griffin. He complains that he is 'a wrapped-up mystery, a swathed and bandaged caricature of a man!'.⁴³ Although he is able to 'pass' as visible when disguised in theatrical costume and make-up, Griffin's body remains insubstantial in comparison to the clothes he wears. He is less a human being than a mystery for the other characters in the text to unravel. Like numerous nineteenth-century depictions of fallen women, where 'clothing, its patterns and its textures, dominates and at times erases the bodies beneath it', Griffin is repeatedly described not as a person, but as his clothing.⁴⁴ Mrs Hall, the landlady of the Coach and Horses Inn where Griffin lodges, views him as 'a brown gloved hand', 'inscrutable blue glasses', and 'a dripping hat brim'.⁴⁵ This is first account we are given of Griffin, and other characters in the village similarly describe him as 'the stranger, muffled in hat, coat, gloves and wrapper'.⁴⁶ Even Kemp looks on him as 'the devouring dressing gown'.⁴⁷ Having desired invisibility in order to escape the problems of his relative poverty, Griffin finds that he is no longer treated as a social subject; he is only a physical body whose vulnerability must somehow be negotiated. Unlike the Angel of *The Wonderful Visit*, who is literally 'fallen' from the fourth dimension, Griffin is metaphorically 'fallen' because he is unable to transcend his material needs.

Like Churton, Griffin revolts against his newly developed awareness of his material body and its impact on others. He is ridiculed by the inhabitants of Iping, the small Sussex village where he lives after fleeing London, and when he reveals his invisibility to them, the scene is at first horrific:

> 'You don't understand', he said, 'who I am or what I am. I'll show you. By Heaven! I'll show you.' Then he put his open palm over his face and withdrew it. The centre of his face became a black cavity. 'Here,' he said. He stepped forward and handed Mrs. Hall something which she [...] dropped [...] and staggered back. The nose – it was the stranger's nose! pink and shining – rolled on the floor.
>
> Then he removed his spectacles, and every one in the bar gasped. He took off his hat, and with a violent gesture tore at his whiskers and bandages. For a moment they resisted him. A flash of horrible anticipation passed through the bar.
>
> It was worse than anything. Mrs. Hall, standing open-mouthed and horror-struck, shrieked at what she saw and made for the door of the house. Every one began to move. They were prepared for scars, disfigurements, tangible horrors, but *nothing*!⁴⁸

---

42 Wells, *The Invisible Man*, 81. Griffin's concern resonates (though likely unconsciously) in the title of Morgan's book, *The Invisible Man: A Self-Help Guide for Men with Eating Disorders, Compulsive Exercise and Bigorexia*.
43 Wells, *The Invisible Man*, 121.
44 Michie, *The Flesh*, 77.
45 Wells, *The Invisible Man*, 7.
46 Ibid., 16.
47 Ibid., 81.
48 Ibid. 37, original emphasis.

However, shock and awe of the situation quickly degenerates into low comedy, with the rush to exit the inn resulting in a village-wide pile-up:

> People down the village heard shouts and shrieks, and looking up the street saw the Coach and Horses violently firing out its humanity. They saw Mrs. Hall fall down and Mr. Teddy Henfrey jump to avoid tumbling over her, and then they heard the frightful screams of Millie, who, emerging suddenly from the kitchen at the noise of the tumult, had come upon the headless stranger from behind.
>
> Forthwith every one all down the street, the sweetstuff seller, the coco-nut-shy proprietor and his assistant, the swing man, little boys and girls, rustic dandies, smart wenches, smocked elders and aproned gipses, began running […].
>
> [Then] a little procession […] was marching very resolutely towards the house, – first Mr. Hall, very red and determined, then Mr. Bobby Jaffers, the village constable […].
>
> 'Ed or no 'ed', said Jaffers 'I got to 'rest en, and 'rest en I *will*'.[49]

Although able to cope with visible disfigurement, the villagers are first stunned at the appearance of *nothing*.[50] However, the rapid shift to slapstick reduces Griffin's malevolent act of unveiling from sublime to ridiculous.

Just as the concern with dressing and painting is gendered as feminine, unveiling oneself is not a traditionally masculine activity. Ludmilla Jordanova examines the politics of veiling and unveiling, noting that in contrast to the erotic and/or intimidating act of female unveiling, 'the idea of unveiling men is comic, implausible or unthreatening'.[51] According to Jordanova, this is 'possibly because neither mystery nor modesty are male preserves but attributes of the other', at least in heteronormative cultures.[52] However, Griffin is a mystery; he is described as such frequently by the villagers and the narrator. What is initially horrifying about the unveiled Griffin is the fact that, like the Freudian reaction to female genitalia, there is *nothing there to see*; however, this apparent emasculation of Griffin quickly turns the villagers' sense of horror to one of ridicule.

Emasculated and reduced to a void, Griffin is faced with the kind of treatment that patriarchal Churton imposes upon Stella. After recovering Stella from the fraudulent Spiritualist, Churton fantasizes about marrying her. This is the first time he 'sees' Stella, whose physical form is signified by her dress, hat, veil and gloves: 'She looked entrancingly pretty. Those little gloves, how charming to put a ring on the finger beneath – if –'.[53] Likewise, unable to comprehend or contain Griffin, the village constable wishes to bind him with handcuffs under the law. After this final insult, Griffin shifts into hypermasculine behaviour, 'smiting and overthrowing for the mere satisfaction of hurting'.[54] On the run from the angry villagers, and betrayed by Kemp to the police, Griffin decides to

---

49 Ibid., 37, original emphasis.
50 The emphasis on *nothing* calls to mind Freud's theory of the male 'castration complex'. See Freud, *The Basic Writings*, 595.
51 Jordanova, *Sexual Visions*, 96.
52 Ibid., 110.
53 Hinton, *Stella, Scientific Romances*, 71.
54 Wells, *The Invisible Man*, 60.

institute a 'Reign of Terror' over the region.⁵⁵ This reign ends in Griffin's violent death, and 'so ends the [...] strange and evil experiment of the Invisible Man'.⁵⁶ Like Graham's experiment with removing Stella from a social economy that automatically positioned her as other, Griffin's experiment, which inversely removed him from the subject position and turned him into a spectacular object, ends in failure. Otherness – coded as invisibility in *Stella* as well as in *The Invisible Man* – serves 'as medium for deep investigations of the nature of self-awareness'.⁵⁷

'An "Habeas Corpus" of an Uncanny Source'

Griffin's newly visible corpse at the end of *The Invisible Man* is literally, to borrow Linda Cartwright's description of contemporary response to X-rays, a 'manifestation of anxieties about sexual difference, mutilation, and death':⁵⁸

> Everyone saw, faint and transparent as though it was made of glass, so that veins and arteries and bones and nerves could be distinguished, the outline of a hand, a hand limp and prone. [...] And so, slowly, beginning at his hands and feet and creeping along his limbs to the vital centres of his body, that strange change continued. It was like the slow spreading of a poison. First came the little white nerves, a hazy grey sketch of a limp, then the glassy bones and intricate arteries, then the flesh and skin, first a faint fogginess, and then growing rapidly dense and opaque. Presently they could see his crushed chest and shoulders, and the dim outline of his drawn and battered features.⁵⁹

This eerie description of the 'strange change' of Griffin's naked corpse, from his hands to the centre of his body, is strikingly similar to images that were being produced and published at the time Wells was composing the novella, such as the one from 1896 in Figure 5.2.⁶⁰

Keith Williams, Laura Marcus and others have observed that *The Invisible Man* adapted particularly well to film in James Whale's Universal Studios version (1933) because of the inherently cinematic qualities of this text, particularly 'the play of absence and presence, and "the presence of an absence", central to theorisations of filmic technology'.⁶¹ While this is undoubtedly an insightful observation, it overlooks Wells's more immediate engagement with another technology of representing 'the presence of an absence', the X-ray radiograph.

One key difference between the representation of human invisibility in Wells and Hinton is Wells's emphasis on the visual uncanniness of the transparent human body, and on the traumatic process of becoming invisible. While this emphasis is partly indicative of Wells's

---

55 Ibid., 125.
56 Ibid., 149.
57 Simpson, 'The "Tangible Antagonist"', 135.
58 Cartwright, *Screening*, 117.
59 Wells, *The Invisible Man*, 148.
60 First angiogram taken by Eduard Haschek and Otto Lindenthal in January 1896. The image appeared in 'X Ray Photography', *Scientific American*, 155.
61 Marcus and Nichols, *Cambridge History of Twentieth-Century Literature*, 338.

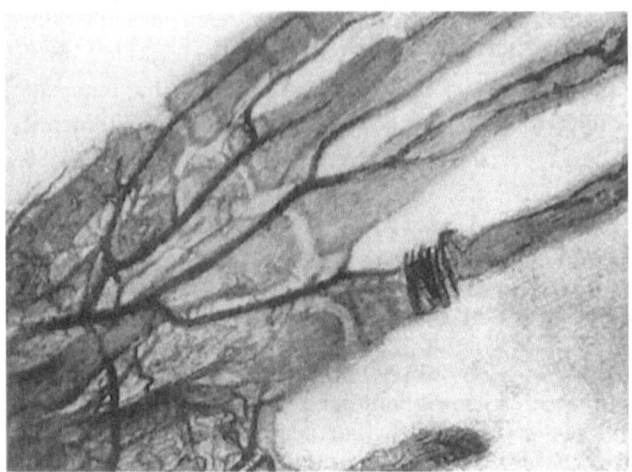

Figure 5.2

flair for the fantastic, it is also the result of the discovery of X-rays in November of 1895. Published in the autumn of 1895, Hinton's *Stella* is unfortunately timed; had he been able to incorporate X-ray imagery into the text, the sales of *Stella* might have been much higher, given the widespread interest in Wilhelm Röntgen's discovery. As it was, Sonnenschein wrote to Hinton in March 1896 that sales for the 1895 edition were unexpectedly low.[62] *Stella* was reprinted later within the second series of *Scientific Romances*; it is likely that Hinton's story was then read as a second-rate imitation of *The Invisible Man*.

Even before the publication of the first X-ray images, Hinton and his contemporaries were particularly concerned with invisible forces and agents: during the second half of the nineteenth century, the invisible provided a conceptual gathering site for various concerns, including those expressed in hyperspace philosophy. 'After the two laws of thermodynamics entered public awareness', Gillian Beer writes, 'the invisible seemed to make us simply receptors of its traffic. [...] By the 1850s, the invisible world might seem to be out of human control'.[63] We have seen how Hinton's invisible Persian king manipulates the valley-dwellers. While Hinton explicitly celebrated the altruistic actions of those who can master the invisible powers of the king, this text also raises the disturbing possibility that human agency is controlled by unseen forces. Whether these forces were conceived as external physical energies, supernatural agents or, perhaps more disturbingly, internalized unconscious dimensions of being, it is not surprising that, in Beer's words, 'tussles developed for the control of meaning relating to that which is invisible'.[64] Nothing served as a lightning rod for this conflict like the discovery of 'a new kind of light', a 'ray' which emanated from an unknown source (hence the 'X'), and could visually penetrate the living flesh of the human body.

---

62 Sonnenschein to Hinton, 10 March 1896, The Records of George Allen & Unwin, Ltd., MSS 383, Book 27, 956.
63 Beer, 'Authentic Tidings', 87.
64 Ibid., 85.

In December 1895, Röntgen published a paper titled 'Über eine neue Art von Strahlen', in the *Proceedings of the Würzberg Physical Medical Society*. In this paper he announced his discovery of the X-rays, and he sent copies, along with initial X-ray images, to a number of recognized physicists in Germany, France and England for evaluation. These images were leaked to Vienna's *Die Presse*, which published a report on 5 January 1896. The following day the London *Daily Chronicle* picked up the story.[65] Ten days later, a brief paragraph and copies of Röntgen's original X-ray images appeared in *Nature*:

> Prof. W. C. Röntgen [...] is reported to have discovered that a number of substances which are opaque to visible rays of light, are transparent to certain waves capable of affecting a photographic plate. It is alleged that he has been able to utilise his discovery to photograph metals enclosed in wooden or woollen coverings, and has succeeded in obtaining pictures showing only the bones of living persons.[66]

The tone here is cautious and sceptical: Röntgen is *reported* to have discovered the rays, and it is *alleged* that he can photograph a living person's bones. This would quickly change, as a barrage of letters, articles, X-ray images and book notices flooded the pages of *Nature* as well as other scientific and popular periodicals. Linda Dalrymple Henderson notes that in 1896 alone, 'more than fifty books and pamphlets and well over a thousand papers were published on the subject of x rays'.[67] Röntgen's choice of the human body as one of the first objects for X-ray imaging was crucial, as a contributor to *The Nineteenth Century* observed early in 1896:

> The wonderful photographs of the bones within the living human body by the Würzberg professor [...], as well as the mysterious character itself of 'invisible rays of light which reveal things concealed from the human eye,' have contributed a great deal to render the discovery widely popular.[68]

The relative accessibility of the equipment needed to generate the images was another contributing factor to 'X-ray mania'.[69] The proliferation of X-ray images and the fast pace of development of X-ray technology are apparent from the correspondence pages of *Nature* throughout 1896 – pages dotted with contributions from enthusiasts who passed along photographs clipped from foreign journals, as well as the results of their own amateur experimentation. The majority of these images are of human hands, similar to Röntgen's first image, which was of his wife's hand. The figure of the skeletal hand became a central motif in early X-ray photography. Thomas Mann depicted the sense of

---

65 Glasser, *Dr. W. C. Röntgen*, 54.
66 'Note' *Nature* 53: 253.
67 L. Henderson, 'X Rays', 324.
68 'Recent Science', *Nineteenth Century* 39: 416.
69 See 'Note' in *Nature* (20 February 1896): 308, where the contributor writes that 'Wm. Wallace and H. C. Pocklington in the Physical Laboratory of the Leeds Central Higher Grade School' obtained X-ray images using 'a cheap incandescent lamp of low candle-power [...] in place of a Crookes tube'.

fascination that such images inspired during the early days of X-ray imaging in *The Magic Mountain* (1924):

> Hans Castorp saw, precisely what he must have expected, but what is hardly permitted man to see, and what he had never thought it would be vouchsafed him to see: he looked into his own grave. The process of decay was forestalled by the powers of the light-ray, the flesh in which he walked disintegrated, annihilated, dissolved in vacant mist, and there within it was the finely turned skeleton of his own hand, the seal ring from his grandfather hanging loose and black on the joint of his ring-finger. [...] He gazed at this familiar part of his own body, and for the first time in his life he understood he would die.[70]

Castorp's unsettling sensation is perhaps best described in the language of the uncanny: the X-rays subvert not only traditional notions of interior and exterior space, thus instigating a 'return of the repressed', but also – by showing the body prematurely stripped bare of its most perishable tissues – they confuse the boundary between the living and the dead. The doctor who is X-raying Castorp here sums up the effect aptly: 'Spooky, what? Yes, there's something definitely spooky about it'.[71]

The X-rays were disturbing also because, as Victorian physicist Arthur Schuster observed, they 'upset all of one's notions of the laws of nature'.[72] The transgressive power of X-rays was not just limited to material notions of nature; human 'nature' was called into question as well. Stripped of their flesh, the X-rayed subjects were nearly indistinguishable: race, class and even gender were not immediately apparent. 'At stake' here, Lisa Cartwright observes, 'is the loss of the cultural text inscribed in the skin, the organs'. For many, 'desire depends on the presence of a surface that conceals a living structure, a signifying surface of clothing or skin that can be read for signs of sexual and cultural difference', hence the problem of invisible subjects such as Stella and Griffin.[73]

As a woman, and thus an object of desire for Churton, Stella is under constant pressure to wear something, or at least 'paint' herself so that she will be visible to him. Griffin's masculine subjectivity is thrown into crisis as a result of the necessity of his continual regard for personal appearance, which – as indicated in Graham's philosophy – is the instigator of feminine identity development. Similarly, Hans Castorp's encounter with the X-ray cited above is fraught with anxiety about submitting himself to the male doctor's gaze. In the waiting room, Castorp watches another patient, a woman, who is also waiting to be X-rayed. While imagining sharing the doctor's visual access to her body (the female patient also models for the doctor, who is an amateur painter), Castorp's 'voyeuristic fantasy is abruptly checked' when he realizes that he too will be reduced to an aesthetic object under the doctor's all-penetrating gaze.[74]

In violating the boundaries of the subject and rendering that subject into spectacular object, the X-rays posed a threat to masculine subjectivity. This was managed in the

---

70 Mann, *The Magic Mountain*, 218–19.
71 Ibid., 219.
72 Schuster, 'Letters', 268.
73 Cartwright, *Screening*, 119.
74 Ibid., 123. See also 123–25, where Cartwright offers a close reading of this scene.

early days of X-ray imaging by assimilating its powers into the pre-existing specular economy: the majority of the early human subjects of X-ray photographs were female (including the first, Bertha Röntgen):

> Among the many physicians who immediately repeated Roentgen's experiments, a woman's hand, sometimes captioned as 'a lady's hand,' or 'a living hand,' became the popular test object. [...] In the public sphere as in medicine, the female hand X ray became a fetish object par excellence. In an early discussion of fetishism, Freud refers to the practice of foot binding, wherein the foot is mutilated and then venerated as an icon of timeless feminine beauty. This image suggested that the female body, like the fetish object, is ownable (as the wedding band, visible in the X-ray image [of Bertha Röntgen's hand], seems to suggest).[75]

Early anxieties about X-ray imaging were not, as we would perhaps expect, about radiation and tissue damage, but rather the disintegration of the cultural signifiers of social identity, as well as the uncanny power they afforded to medical men who made use of them. Cartwright demonstrates how the discovery of X-rays 'was received with such widespread excitement [...] because it further legitimated [the] model of visual knowledge as corporeal penetration and invasion, a model that previously had currency as popular fantasy and spectacle'.[76] However, numerous early responses to this discovery also indicated anxiety that the boundary between public and private might be completely dissolved.

A poem in the 1897 volume of *International Annual of Anthony's Photographic Bulletin and American Process Yearbook* (published in 1896) brings together a number of concerns about the new photography. The male persona of the poem warns:

> Your sweetheart your photo will not accept,
> "'T is too old-fashioned by half."
> Though you deemed the portrait beyond compare,
> She says in her locket she prefers to wear
> Your skeleton radiograph.[77]

In addition to the request that he be 'penetrated' by his lover's gaze, the implied male reader is told he can look forward to his 'inmost perception' – his thoughts – being revealed. The poet, Emily Culverhouse, footnotes an article in the *Standard* to support the claim that 'thought photography' is possible, and then continues to describe a scenario:

> I say to my wife, 'Don't wait supper for me,
> For my business may cause a delay.'
> But the Röntgen rays enlighten her eyes,
> And by their deep aid she quickly decries
> 'Tis pleasure that stands in the way.

---

75 Ibid., 115. See also Tovino, 'Imaging'.
76 Cartwright, *Screening*, 114.
77 Culverhouse, 'Photography', 75–78.

> She sees in my pocket, so snugly ensconced,
> A bracelet in case, and a fan;
> And doubtless this all-searching thought-reading too,
> Brings name and appointment at once into view,
> So she frustrates my nice little plan.[78]

Even men are subject to the penetrating gaze of the X-rays, as this seemingly omniscient power is accessible to women as well. The poem demonstrates the downside of Hinton's proto-X-ray, four-dimensional vision in which 'our secrets lie as clear as the secrets of a plane being lie to an eye above the plane. [...] And so we lie palpable, open. There is no such thing as secrecy'.[79] Culverhouse's doggerel verse describes a future where 'we shall live with our friends, and our relatives too / In a state of mistrust and suspicion', and 'an Englishman's body' no longer 'belongs to himself'.[80] After offering the spectre of men, as well as women, being subject to this panoptic gaze, Culverhouse concludes:

> If our houses are raided, the law will step in;
> Then in justice and plain common sense,
> Our bodies and minds should receive the same aid;
> And to pry without warrant should surely be made
> An illegal and heinous offence.[81]

This is 'an "habeus corpus" of an uncanny source', Culverhouse argued, reminding the reader that X-rays are 'a gruesome, weird, and mystical force / (But clothed in the garb of science of course)'.[82] From the very beginning, X-rays were associated with the occult. In his biography of Röntgen, Otto Glasser suggested that Röntgen's initial investigations were carried out in secret because, after first witnessing the uncanny penetrative power of the X-rays, he feared being brought into 'disrepute in the eyes of his colleagues' if it became known that he was conducting metaphysical experiments.[83]

The X-rays added another layer to the 'garb of science' around the fourth dimension as well. As early as 'What Is the Fourth Dimension?', Hinton described the kind of vision that the X-rays seemed allow:

> A being in three dimensions, looking down on a [two-dimensional] square, sees each part of it extended before him, and can touch each part without having to pass through the surrounding parts, for he can go from above, while surrounding parts surround the part

---

78 Ibid., 77.
79 Hinton, 'Many Dimensions', 42.
80 Culverhouse, 'Photography', 77 and 75.
81 Ibid., 78.
82 Ibid., 75.
83 Glasser, *Dr. W. C. Röntgen*, 38. See also Cartrwright, *Screening*, 114–15.

he touches only in one plane. *So a being in four dimensions could look at [...] every part of a solid figure.*[84]

Hinton recognized the potential of this vision for medical science; in the 'sexual inversion' anecdote from 'A Plane World', Mulier returns to her two-dimensional space not only with male sexual characteristics, but also 'strange knowledge of the internal anatomy' of her species, and 'most of their medical knowledge dates from her'.[85]

While Hinton never explicitly linked X-rays with four-dimensional vision, many others did. Second-generation hyperspace philosopher Claude Bragdon argued that 'with the evidence of X-rays it could be regarded as even more scientifically respectable to believe in the demonstrable evidence of a condition of "four-dimensional vision" which rendered material objects transparent'.[86] Wells played with this newfound respectability to make his invisible man plausible; Griffin's explanation of the secret of invisibility links four-dimensional theory with X-rays. After discovering the four-dimensional principle of pigments and refraction, Griffin tells Kemp that

> the essential phase was to place the transparent object whose refractive index was to be lowered between two radiating centres of a sort of ethereal vibration, of which I will tell you more later. No, not these Röntgen vibrations – I don't know that these others of mine have been described.[87]

Thus, Wells was able to capitalize on Röntgen's fantastic discovery in a way that Hinton, in *Stella*, was not.

Wells's Splintering Frame Technique and 'Cubist Visual Culture'

In his examination of Wells's work in the twentieth century, Scheick argues that in his later fiction, Wells used a

> technique of 'the splintering frame' [that] employs certain fictional conventions in a way designed [...] to frustrate reader expectations aroused by these conventions, [...] to draw attention to the artificiality and ideology behind these conventions, and [...] to point away from the 'exhausted' text as self-contained, finished artifact and towards the self-aware reader, who ideally participates within the expanded boundary of the text and discovers within himself [*sic*] a capacity (dimensionality) for a heightened awareness of and control over human fate. This technique of the splintering frame comprises the aesthetic fourth dimension of his novels.[88]

Scheick overlooks Wells's early fiction, but even here Wells was already experimenting in the literary fourth dimension; particularly in *The Invisible Man*, with its emphasis on

---

84 Hinton, 'What Is the Fourth Dimension?', *Scientific Romances*, 13, emphasis added.
85 Hinton, 'A Plane World', *Scientific Romances*, 147.
86 Bragdon, *Four-Dimensional Vistas*, 57.
87 Wells, *The Invisible Man*, 95.
88 Scheick, *The Splintering Frame*, 24–25.

transgressive developments in visual technology, we can observe early examples of his 'splintering frame technique'.

To splinter the narrative frame of a written text is to enact a kind of literary *trompe-l'oeil* which – as in Picasso and Braque's collage work – through deception of the eye, the *mind* is 'undeceived'.[89] Wells believed this 'undeception' of literary realism was necessary at the end of the nineteenth century:

> Throughout the broad smooth flow of nineteenth-century life in Great Britain, the art of fiction floated on this [...] assumption of social fixity. The Novel in English was produced in an atmosphere of security for the entertainment of secure people [...]. Its standards were established within that apparently permanent frame and the criticism of it began to be irritated and perplexed when, through a new instability, the splintering frame began to get into the picture.[90]

Wells was less interested in developing a refined mimesis of 'real life' than he was with disrupting what he viewed as the 'broad smooth flow' of nineteenth-century realism. He disrupted this realism by introducing fantastic and metatextual elements into his fictions. Baudrillard is useful here:

> Trompe-l'oeil does not attempt to confuse itself with the real. Fully aware of play and artifice, it produces a simulacrum by mimicking the third dimension, questioning the reality of the third dimension, and by mimicking and surpassing the effect of the real, radically questioning the principle of reality.[91]

Wells's narratives were designed to function by analogy: by splintering the mimetic frame of the text, he encouraged the movement of the reader's attention from the world represented within the text to events in the external world of social and lived reality. Like the visitation of the three-dimensional Sphere that ruptures A. Square's two-dimensional perceptual frame in *Flatland*, Wells disrupted the narrative frame to emphasize to the reader the constructed nature of the 'real' world beyond the literary text.

This was a political and ethical endeavour, a call to his readers to take action. In 1911, Wells declared the novel 'a powerful instrument of moral suggestion'; it was, in fact, 'the only medium through which we can discuss the great majority of the problems which are being raised in such bristling multitude by our contemporary social development'.[92] He appears to have had some success in achieving this goal of increasing his readers' social awareness and political engagement, if we consider Virginia Woolf's complaint that Wells's stories 'leave one with so strange a feeling of incompleteness and dissatisfaction. In order to complete them it seems necessary to do something – to join a society, or [...] to write a cheque'. The Wellsian novel is not even

---

89 See Greenberg, *Art and Culture*, especially 72.
90 Wells, *Experiment*, 1: 494–95.
91 Baudrillard, *Selected Writings*, 156.
92 Edel and Ray, eds, *Henry James and H. G. Wells*, 144 and 148.

a novel by Woolf's standards ('Sometimes I wonder if we are right to call them books at all'); it is simply a text held together by its physical binding rather than its aesthetic unity.[93] However, this unfinished, outward-looking quality was crafted intentionally: it was Wells's literary aesthetic, a way of arousing a higher-dimensional consciousness in the reader.

Wells believed that the turn of the twentieth century was a particularly opportune time to reconsider the art and scope of the novel because the cumulative effects of the social, cultural and technological upheavals of the second half of the nineteenth century had resulted in a 'great intellectual revolution [...] of which the revival and restatement of nominalism under the name of pragmatism is the philosophical aspect'.[94] As Hinton and even Stewart and Tait observed earlier, with the decline of religious certainty and the increase of cosmopolitanism, the 'feeling of certitude about moral values and standards of conduct' was fast dissipating.[95] In this context, the fourth dimension can be read as an attempt to negotiate the aporia created by this cultural shift. In the case of Stewart and Tait, this is certainly the function it fulfilled: the unseen spiritual universe that exists somewhere in the fourth dimension was the means by which the immortality of the soul (and therefore the need for moral absolutes in mortal life) could be preserved in the face of mounting scientific evidence to the contrary.

Hinton's conception of the fourth dimension differed from Stewart and Tait's: it was not a stand-in for a metaphysical centre, or a universalizing moral agent. Hinton's hyperspace was accessible by a 'moving consciousness', one that would vary from individual to individual; moral value is similarly relative here. Wells's four-dimensional aesthetic is consonant with Hinton in opposition to Stewart and Tait. For Wells, post-Victorian morality was to be discursive, and the novel was the most apt vehicle for exploring this shift: 'I do not mean merely that the novel is unavoidably charged with the representation of this wide and wonderful conflict. It is a necessary part of the conflict'.[96] 'Ideally', Scheick writes, for Wells, the outcome of the novel's involvement in this conflict would result in 'both reformed [narrative] structure and reader sensibility becom[ing] four dimensional or, in other words, open to ever-expanding indeterminate possibilities'.[97]

While Wells's experiments in reforming narrative structure led him down various paths, we can find an early – and quite literal – example of his splintering frame technique in an 1894 short story, 'Through a Window'. In this narrative, a man named Bailey is convalescing in his home in West London. With two broken legs, he is unable to move from the couch of his study; his sole entertainment is the view from his window, which looks out onto the Thames. 'Funny', he remarks, 'how these people [on the river] come from all points of the compass – from Oxford and Windsor, from Asia and Africa – and

---

93 Woolf, *The Captain's Death-Bed*, 99.
94 Edel and Ray, eds, *Henry James and H. G. Wells*, 147–48.
95 Ibid., 145.
96 Ibid., 147.
97 Scheick, *The Splintering Frame*, 25.

gather and pass opposite the window just to entertain me'.[98] However, one day the man's self-centred worldview is disrupted when this 'entertainment' violently intrudes into his study in the form of a hunted escapee, a Malay servant from one of the pleasure boats in the area. The man is shot by his pursuers as he is climbing through the window frame into Bailey's study: he falls dead on top of Bailey, 'rapidly staining and soaking the spotless bandages' of the invalid.[99] Here, disturbing elements of the external world – external to Victorian household and empire – force themselves into the comfortable sphere of the bourgeois audience, and by 'staining' it, alter it irreversibly. Bailey, the passive and 'secure' consumer of the narrative circumscribed by his window frame, is shocked out of complacency when that frame, the boundary between real and fictitious space, is abruptly violated in a scene that recalls the boy's transgression in Pere Borrell del Caso's *Escapando de la Crítica* (1874). While the subject of del Caso's painting may be attempting to escape the critical gaze, it is Bailey – the stand-in for the reader – who is unable to escape in Wells's narrative.

Ten years later, Wells shifted away from dramatizing the splintering frame toward utilizing it as a structural effect in *A Modern Utopia* (1905). Here Wells constructed an elaborate frame for the narrative, first addressing the reader as himself, in 'A Note to the Reader', and then with another introduction titled 'The Owner of the Voice', written entirely in italics. '*Now this Voice*', according to Wells, '*is not to be taken as the Voice of the ostensible author who fathers these pages. You have to clear your mind of any preconceptions in that respect*'. After a detailed description of this narrative voice and the physical person from whom it issues, the text continues:

> *Him you must imagine as sitting at a table reading a manuscript about Utopias [...]. The curtain rises upon him so. But afterwards, if the devices of this declining art of literature prevail, you will go with him through curious and interesting experiences. Yet, ever and again, you will find him back at the little table, the manuscript in his hand, and the expansion of his ratiocinations about Utopia conscientiously resumed.*[100]

Foregrounded here is the movement that Wells wanted his reader to mimic, from the narrative of the inner text to the 'outside' world of the narrator. 'This declining art of literature' is the fantasy-making activity that creates the Utopia, but it is also exposed here as an 'art', as artifice. Another, more 'metafictional' example of splintering the frame occurs within the internal narrative frame of *Boon*, to which I turn shortly. First, I want to consider Wells's experimentation with the structure of *The Invisible Man* as an early attempt to splinter the frame. This text resonates with allusions to the X-ray's ability to confuse internal and external spaces, and Wells's form matches his content; there are frequent shifts in voice and style that abruptly disrupt reader expectations. The reader's attention is thus 'raised' from the superficial conventions of nineteenth-century realism to a higher dimension of self-awareness.

---

98 Wells, *The Complete Short Stories*, 31.
99 Ibid., 35.
100 Wells, *A Modern Utopia*, 7.

Disruption and Disunity in *The Invisible Man*

Wells dramatized the hermeneutic process in the first part of this text by depicting the villagers in the act of 'reading' Griffin; later in the text, Wells's experiments with the narrative structure foreground the act of reading itself. *The Invisible Man* 'unfolds' in two senses of the word: there is quite literally an unravelling of the layers of representation surrounding Griffin before he is finally rendered visible again as a naked, battered corpse. Before the readers – both inside and outside the narrative frame – are allowed to 'see' his body, they must first move through multiple and sometimes contradictory perspectives on Griffin. This shifting viewpoint of the eponymous character mimics Hinton's textual constructions of the fourth dimension throughout his *Scientific Romances*. Just as Hinton asked, 'What is the Fourth Dimension?' in his first romance and then proceeded to offer multiple answers, Wells opened *The Invisible Man* with a chapter titled 'The Strange Man's Arrival', and proceeded to construct various answers to the implied question regarding this man's identity.

First the reader is given the observations of the innkeeper, Mrs Hall; then the narrative expands to include the impressions and speculations of the other residents of the village of Iping. The events of the narrative and the developing hypotheses of the villagers are presented in the matter-of-fact tone of a case-study. As Steven McLean observes, 'the influence of the detective genre is apparent as the characters make a series of increasingly sophisticated guesses as to the identity concealed by the stranger's bandaged condition'.[101] The villagers progress from the assumption, based on Griffin's bandages, that he has suffered a disfiguring accident, to speculations that he is hiding from the police, a prospective burglar, that his skin is black, or 'he's a half-breed', a 'harmless lunatic' or somehow supernatural.[102] Once faced with the truth of Griffin's situation, the characters constantly question their abilities to perceive and understand reality: 'Am I mad?' Cuss began abruptly [...]. 'Do I look like an insane person?'; 'Am I drunk?' said Mr. Marvel. 'Have I had visions?'; and 'Kemp slapped his brow with his hand. "Am I dreaming? Has the world gone mad – or have I?"'[103]

Churton's worry that if he left Stella to the care of the Cornish family 'you would all be persuaded again that Stella doesn't exist' is confirmed by the villagers' reaction to Griffin. When faced with the seemingly impossible existence of an invisible man, they soon convince themselves of his unreality: 'After the first gusty panic had spent itself Iping became argumentative. Scepticism suddenly reared its head [...]. It is so much easier not to believe in an invisible man'.[104] On many levels, Griffin simply does not fit into the villagers' paradigm for perceiving and understanding the phenomenal world; however, frustrated by their inability to 'read' him accurately, Griffin refuses to be ignored. As Anne B. Simpson observes, 'that the townspeople themselves are [morally] bankrupt is illustrated by their notably limited discourse on Otherness'.[105] Refusing to undertake

---

101 McLean, *The Early Fiction*, 69.
102 Wells, *The Invisible Man*, 20 and 23.
103 Ibid., 24, 44 and 85.
104 Ibid., 49.
105 Simpson, 'The "Tangible Antagonist"', 136.

a will-to-attend in order to confront something outside of their perceptual frame, they instead first try to ignore Griffin, imposing a kind of social death; when this fails, they must actually kill him.

The early chapters of *The Invisible Man* present the villagers' attempts to uncover the reason for Griffin's bandages and strange behaviour, but after he reveals his invisible body to them the narrative shifts from detective mode to low comedy and broad Biblical parody.[106] When Wells introduces Dr Kemp, a scientist and former acquaintance from Griffin's student days, the tone shifts again into the first-person singular as Griffin relays his version of events to Kemp. It is only at this point, precisely midway through the text, that we are given Griffin's name. Before his conversation with Kemp, Griffin is referred to as 'the Invisible Man', 'the stranger', the 'Bogey Man', 'the Unseen', 'the voice' and various items of clothing.

As Griffin describes the struggles of his past as an impoverished student and scientific researcher living in a London slum, and the difficulties brought about by his invisibility – he is perpetually ill from walking around naked in the cold; he is harassed by children, dogs and the blind; and he risks being run down every time he ventures onto the street – he becomes more of a sympathetic figure. However, Wells undercuts this first-person narration by embedding it within a larger dialogue between Kemp and Griffin. Kemp works as a foil to Griffin's narration, questioning and protesting at Griffin's actions. He challenges Griffin when he tells of how he spitefully set fire to his boarding house in London, and by his feeble justification, that 'it was the only way to cover my trail – and no doubt it was insured', Griffin again reveals his violent temper.[107] Any straightforward reading of Griffin's villainy is further complicated when he anxiously implores Kemp: 'You don't blame me, do you? You don't blame me?'[108]

Numerous critics regard the ambiguous nature of Griffin's character as a significant technical flaw. John Batchelor argues that 'the reader's sympathies for much of the novel are *for* Griffin and against, for example, the oafish inhabitants of Iping [...]. Half-way through this novel Wells forces it against the grain of our sympathies – and severely damages it in my view'.[109] In Batchelor's opinion, *The Invisible Man* is one of Wells's 'less successful' romances because the question, 'Is the novel finally a moral allegory about the abuse of science or a heroic fable about an outsider who refuses to live by middle-class standards?' remains unanswered.[110] It is this kind of reductive reading that Wells challenged in the elaborate framing of *A Modern Utopia*. Wells's frustration with emphasis on unity of aesthetic vision was also at the heart of his quarrel with Henry James.

In opposition to Batchelor, Robert Sirabian claims that the disunity of *The Invisible Man* allows for 'an explorations of tensions' of opposing worldviews, making this text 'more than a scientific romance'.[111] However, the foregrounding of such tensions *is* in

---

106 It is beyond the scope of this discussion to fully address the elements of Biblical parody in *The Invisible Man*; see instead Stetz, 'Visible and Invisible Ills'.
107 Wells, *The Invisible Man*, 102.
108 Ibid., 119.
109 Batchelor, *H. G. Wells*, 22, original emphasis.
110 Ibid., 30.
111 Sirabian, 'The Conception', 384.

fact what makes this a scientific romance. The combination of science and fantasy, and the tensions between the romantic subject and strictures of 'objective' reality, underlie the unstable fiction of the fourth dimension in Hinton's *Scientific Romances* and were foregrounded by Wells in his attempt to develop a four-dimensional literary aesthetic. *The Invisible Man* is a 'grotesque romance' not only because of the gruesome descriptions of Griffin's body, but because even though he is a monstrous and morally reprehensible figure, he is not represented as totally inhuman (despite Kemp's claims).[112] In fact, there are places where the narrator encourages us to empathize with Griffin.

After the dialogue between Griffin and Kemp, the narrative is briefly drawn outward to an omniscient, third-person account of events as Kemp betrays Griffin to the police. Once Griffin is on the run again, however, the narrative shifts yet again; in reporting Griffin's attack on a local resident in 'The Wicksteed Murder' chapter, the tone is journalistic. Here the narrator relays the evidence found at the murder scene, oscillating between cool speculation on motive and sensational description: 'The Invisible Man seems to have rushed out of Kemp's house in a state of blind fury. [...] No one knows where he went nor what he did. But one can imagine him hurrying through the hot June afternoon'. The narrator pulls the reader's emotions in opposing directions by arguing sympathetically for Griffin, while also describing the outcome of his violent behaviour in the details of the vicious murder that he apparently perpetrated. Then we are encouraged to empathize with Griffin's outrage at Kemp's betrayal:

> No doubt he was almost ecstatically exasperated by Kemp's treachery, and [...] we may still imagine and even sympathize a little with the fury the attempted surprise must have occasioned. [...] He had evidently counted on Kemp's cooperation in his brutal dream of a terrorized world.[113]

Here the request for the reader's identification is undercut by a reminder of Griffin's desire for a 'Reign of Terror'. The narrator also speculates on the possibility of extenuating circumstances that could lift 'the murder out of the realm of the absolutely wanton', by reminding the reader that Griffin was being pursued and hounded by the police and angry villagers, and that 'the evidence that he had the iron rod [the murder weapon] in hand before he met [Mr] Wicksteed is [...] overwhelming'.[114]

The narrator continues, painting a scenario in which Griffin kills Wicksteed in self-defence, explaining that 'no doubt the Invisible Man could have easily distanced his middle-aged pursuer under ordinary circumstances, but the position in which Wicksteed's body was found suggests that he had the ill luck to drive his quarry into a corner'.[115] Again, the mitigating evidence of the position of the body, and the narrator's generous speculation that 'the sight of his victim, bloody and pitiful at his feet, may have released some long pent fountain of remorse' is undermined by the details of the 'bloody

---

112 The complete original title of the novella is *The Invisible Man: A Grotesque Romance*.
113 Wells, *The Invisible Man*, 129.
114 Ibid., 129–30.
115 Ibid., 132.

and pitiful' state of Wicksteed's corpse, recalling an earlier description of how Griffin 'stopped this quiet man, going quietly home to his midday meal, attacked him, beat down his feeble defences, broke his arm, felled him, and smashed his head to a jelly' with an iron rod.[116] In this chapter we see a particularly well-controlled example of how Wells's writing arouses and then subverts reader expectations.

The terror of the final chase scene, wherein Griffin pursues Kemp into Port Stowe with the intention of murdering him, is complicated by the fact that Kemp ends up running 'in his own person the very race he had watched with such a critical eye from the belvedere study only four days ago'.[117] The previous race had been run by the tramp, Thomas Marvel, who was also attempting to escape Griffin's murderous rage. Kemp's observation of Marvel's flight from Griffin is critical in two senses: his gaze is both detached and condemnatory. Like Bailey, Kemp watches the chase from a 'pleasant little room, with three windows, north, west, and south' where 'there was no offence of peering outsiders', a position of superiority from which he judges the scene below: 'Another of those fools,' said Doctor Kemp. 'Like that ass who ran into me this morning round a corner with his "'Visible Man a-coming, sir!" I can't imagine what possesses people'.[118] Kemp's disdain of Marvel's plight is repaid by his own neighbour's selfish reaction to the sight of Kemp later fleeing Griffin:

> He ran to shut the French windows that opened on the veranda; as he did so Kemp's head and shoulders and knee appeared over the edge of the garden fence. In another moment Kemp had ploughed through the asparagus, and was running across the tennis lawn to the house.
>
> 'You can't come in,' said [Kemp's neighbour,] Mr. Heelas, shutting the bolts. 'I'm very sorry if he's after you, but you can't come in!'[119]

Kemp's recapitulation of Marvel's race takes place in a chapter appropriately titled, 'The Hunter Hunted', where we see a number of reversals. Kemp, who has instigated the police hunt for Griffin is now being pursued by Griffin to be his first execution in his Reign of Terror. By the end of this chapter, however, the power has changed hands yet again as residents of Port Stowe chase Griffin and beat him to death in a frenzy of mob violence. After Griffin's corpse returns to visibility, the imagery Wells uses to describe his body is both pathetic and fierce:

> When at last the crowd made way [...] there lay, naked and pitiful on the ground, the bruised and battered body of a young man about thirty. His hair and beard were white, – not grey with age, but white with the whiteness of albinism, and his eyes were like garnets. His hands were clenched, his eyes wide open, and his expression was one of anger and dismay.[120]

Griffin's pre-invisibility 'otherness' is depicted here: he was an albino. Though 'necessary' for the plot (as Griffin explains, it is impossible to render the pigment in hair

---

116 Ibid., 132, and 131.
117 Ibid., 145.
118 Ibid., 70.
119 Ibid., 144.
120 Ibid., 148.

transparent), with his albinism, Griffin is already an exception to Anglo-Saxon normativity. Even before his body becomes a site for a confluence of anxieties from the supernatural to the transgressive power of X-rays, his pigmentless body is a sort of parody of the discourse of 'whiteness'. Bettyann Holtzmann Kevles cites two separate occasions where American doctors reported that they 'had successfully combined X-ray treatment with radium to bleach the skins of Negros white'; Griffin, like the subjects of these appalling experiments, is an 'other' who is transformed into a caricature of normativity.[121]

Griffin's downfall is precipitated by the fact that though his body is invisible as a spirit, he remains crudely physical. Throughout the text he is circumscribed by material concerns: lack of money and privacy, hunger and the vulnerability of his naked, transparent body. Within the narrative frame of *The Invisible Man*, Griffin is a fiction, a fantasy of omniscient masculinity – Ruskin's paradoxical invisible man – brought grotesquely to life. Though his invisibility gives him power over his contemporaries, he is still reliant on others to confirm his identity ('What good is pride of place when you cannot appear there?' he asks Kemp).[122] Even the most educated of the other characters, Dr Kemp, is unable to adequately 'read' Griffin – it is only with Griffin's assistance that Kemp is able to register the events leading up to his invisibility. As a prosaic and superficial reader, Kemp is not able to raise Griffin above the status imposed upon him by the villagers as monstrous and other.

It is near the end of his conversation with Kemp that Griffin begins to 'dream of playing a game against the race'.[123] Kemp responds by urging Griffin: 'Publish your results; take the world – take the nation at least – into your confidence'.[124] The request that Griffin publish his results (and himself – his invisible body *is* the result of his research and experimentation) before the public gaze has little impact on Griffin for two reasons: firstly, Kemp only makes the suggestion as an attempt to stall Griffin while he awaits the arrival of the police and, secondly, because having already proved himself illegible to Kemp and the Iping villagers, Griffin does not have reason to hope for anything better by going public. His access to the four-dimensional formula that allows him to seemingly transcend the social contract does not allow him to transcend his material origins. Positioned as a painful intermediary between the three-dimensional everyman and the four-dimensional higher consciousness, Griffin's name is particularly apt: not only is he, like the gryphon, a fantastical beast, but he is also a 'newcomer', and – in racist nineteenth-century evolutionary logic – a combination of 'higher' and 'lower', of mixed racial origin, one who is perhaps able to 'pass'.[125] Not satisfied with passing as this 'caricature of a man', Griffin disrupts the villagers' frame of reality, just as that other visitor from another realm, the Angel, does in *The Wonderful Visit*. Wells took pains to ensure that

---

121 Kevles, *Naked*, 49.
122 Wells, *The Invisible Man*, 121.
123 See Hardin, 'Ralph Ellison's *Invisible Man*', 99. Hardin explores the implications of Griffin's game against the 'race', observing that 'there is an intriguing convergence of "passing," miscegenation, and homoeroticism within the metaphor of invisibility' (97).
124 Wells, *The Invisible Man*, 125.
125 As Hardin observes, the *OED* defines 'griffin' as a gryphon; as a 'greenhorn' or 'newcomer', particularly with reference to a European recently arrived in India; and, in nineteenth-century Louisiana parlance, a 'mulatto'.

we read neither character as metaphysical: his 'Digression on Angels' (which explains that the Angel is not of the religious type, the kind 'which it must be irreverent to touch') interrupts the narrative flow of *The Wonderful Visit*, and the workings of Griffin's internal organs render him 'grotesquely visible' at times.

In her discussion of nineteenth-century microscopic and X-ray imaging, Cartwright argues that 'a "cubist" visual culture developed in part as a cultural response both to the epistemological instability of human observation and to the sight of the human body'.[126] In writing of a cubist visual culture, Cartwright is careful to explain that she does not refer solely to the artistic movement associated with Picasso and Braque, among others; she is also interested in the role that developments in imaging the human body, as well as moving picture technology, played in 'the formation of a pervasively cubist culture – a culture that reconfigures the bodily interior as an endlessly divisible series of flat surfaces and mobile networks'.[127] The 'cubist culture' that Cartwright identifies here can be extended beyond modernist representations of the body to turn-of-the-century literary aesthetics as well. Hinton's attempts to represent four-dimensional objects and four-dimensional consciousness as a series of three-dimensional 'slices' is part of this cubist culture, as is Wells's splintering frame technique. The 'flattening' tendency foregrounded in these cubist texts, which Cartwright calls 'the abhorrence of dimensional form' – in the case of Wells and Hinton at least – resulted from an attempt to represent a higher dimensionality within a lower-dimensional medium. Thus the 'slices', or individual short texts within Hinton's *Scientific Romances*, like the 'solid' paper of his cube exercises, are, in comparison with his idea of a spatial fourth dimension, 'flat'.

Wells, like Braque and Picasso, used a kind of literary *trompe l'oeil* to 'declare' rather than deny the surface, or textuality, of his medium.[128] In this sense his texts are 'flat'; just as his characters are, as E. M. Forster said, 'flat as a photograph'.[129] Unable to accept the premise of his four-dimensional aesthetic, Wells's flat characters and his splintering frame technique were often disregarded by contemporaries as symptoms of his deficiency as a novelist. For the majority of the years Henry James and Wells were friends, these 'deficiencies' were the frequent topic of discussion between the two men; Henry James sought to 'correct' these problems in the younger writer's prose, and the tensions that resulted eventually erupted into Wells's cruel critique of James in *Boon*.

The Spoils of *Boon*

The complex structure of Boon is reflected in its full title: *Boon, The Mind of the Race, The Wild Asses of the Devil, and The Last Trump: Being a First Selection from the Literary Remains of George Boon, Appropriate to the Times, edited by Reginald Bliss, with an Ambiguous Introduction by H. G. Wells*. In this text Wells not only vented his frustration with Henry James's theory of the art of fiction, but he also took some steps toward establishing a theory of his

---

126 Cartwright, *Screening*, 91.
127 Ibid., 91.
128 See Greenberg, *Art and Culture*, 72.
129 Forster, *Aspects of the Novel*, 99.

own. This text was designed in direct opposition to what Wells perceived as James's literary aesthetic: it is fragmented, discursive and full of caricature. Wells refused to respect James's appeal to literary artists to abstain from 'giving themselves away' by interrupting the narrative with authorial commentary.[130] As a final nail in the coffin of the possibility of James approving of Wells's latest literary offering, Wells's mouthpiece, George Boon, parodies James's writing style in a short story he has written (which is included in *Boon*), titled 'The Spoils of Mr. Blandish'.

Presented as the literary remains of popular Edwardian novelist George Boon, *Boon* is composed of the fragments of Boon's speculations, short stories and cartoons, as well as the recollections of the editor (Reginald Bliss), of Boon and their circle of friends.[131] Throughout the various sections of the book, Boon is planning to write a novel on what he calls the 'Mind of the Race', which is to be an update of W. H. Mallock's *New Republic*, featuring discussions between many contemporary literary celebrities. Henry James, George Moore and Edmund Gosse, among others, wander through the fragments of Boon's novel. Other real literary figures also appear within Bliss's recollections: Ford Madox Ford briefly interrupts Boon's discourse on Henry James from over the garden wall where he is playing badminton with a wholly fictional character named Wilkins. The distinctions between the fictional and the metafictional are further blurred when Wilkins questions Boon (along with another character named Edwin Dodd) on his idea of the 'Mind of the Race':

'All through this book, Boon,' he [Wilkins] began.

'What book?' asked Dodd.

'*This one we are in*. All through this book you keep on at the idea of the Mind of the Race. It is what the book is about; it is its theme'.[132]

In this metatextual moment, Wilkins acknowledges his fictional status within the book and, like Dodd, we are unclear whether it is Boon's book or *Boon* referred to here. The confusion resulting from this narrative interruption causes a moment of 'seizure', which, in Baudrillard's words, 'rebounds on the surrounding world we call "real," revealing to us that "reality" is nothing but a staged world'.[133]

The Mind of the Race itself is proposed as a dimension of consciousness that transcends this staged world. Both Wells's book *Boon*, with its fragmented narratives, and Boon's unfinished book discussed within *Boon*, are about this higher consciousness. By splintering the narrative frame of *Boon* with the insertion of extratextual elements, Wells was modelling the path he pushed his reader to take: he wanted them to recognize something beyond the three-dimensional physical and social world. Just as the

---

130 Matthiessen, ed., *The James Family*, 355.
131 The circle of literary friends in *Boon* is likely based on Wells's own friendship group from the late 1890s to around 1915; this circle included Henry James, Joseph Conrad and Ford Madox Ford. See Delbanco, *Group Portrait*.
132 Wells, *Boon*, 179, emphasis added.
133 Baudrillard, *Selected Writings*, 156.

three-dimensional world of the reader exists outside the textual realm of Wilkins, Dodd and Boon, this higher dimensional consciousness – Boon's Mind of the Race – exists outside of the reader's experience. Boon describes the Mind of the Race as

> something more extensive than individual wills and individual processes of reasoning in mankind, a body of thought, a trend of ideas and purposes, a thing made up of the synthesis of all the individual instances, something more than their algebraic sum, [...] a common Mind expressing the species.[134]

This Mind, according to Boon, was just beginning to awaken to self-consciousness at the turn of the twentieth century. Drawing on the language of Emerson and Nietzsche, Boon/Wells described this Mind of the Race as a fusion and division of 'Over-minds' for which 'there is no birth, no pairing and breeding, no inevitable death. That is the lot of such intermediate experimental creatures as ourselves'.[135] Unlike the *Übermensch*, Wells's Over-minds are *plural*: they are the collective consciousnesses of various human cultures, which in turn combine to form a singular Mind, a universal collective.

Just as Hinton wanted to make his readers aware of a space or a condition of being that transcends the petty differences of everyday life, Boon imagines a transcendent conglomeration of human consciousness – the Mind of the Race – that can look impartially upon the three-dimensional world of individuals. Boon sees literature as one way of awakening this Mind to consciousness: 'Literature, the clearing of minds, the release of minds, the food and guidance of minds, is the way, Literature is illumination, the salvation of ourselves and of every one from isolations'.[136] When, citing past and present atrocities such as the Great War, Wilkins challenges Boon's theory of a collective Mind of the Race, claiming it might simply be 'a gleam of conscious realization that passes from darkness to darkness –'. Boon can only support his useful fiction by enacting his own will-to-believe: '"*No*. [...] Because I will not have it so," said Boon'.[137]

The perspective afforded to this collective mind sounds very similar to the view from that other useful fiction, Hinton's fourth dimension. The rise to consciousness of the Mind of the Race may indeed result in Hinton's 'new era' of thought. Hinton argued that 'to our ordinary space-thought, men are isolated, distinct, in great measure antagonistic. But with [...] higher thought, it is easily seen that all men may really be members of one body, their isolation may be but an affair of limited consciousness'.[138] Boon explains that the rise to self-consciousness of the unified Mind of the Race is necessary for the peaceful progress of humanity; in fact, he argues, the present failure to recognize it lay at the root of the Great War. If literature is the key to the Mind of the Race, then, the implication is that inappropriate forms of literature – such as James's – are actually detrimental to the progress of humanity. Boon is rather unequivocal about the value of

---

134   Wells, *Boon*, 42.
135   Ibid., 44.
136   Ibid., 209.
137   Ibid., 183, original emphasis.
138   Hinton, *A New Era*, 97.

Henry James's aesthetic: When asked by a companion, 'Ought there, in fact, to be Henry James?', Boon responds, 'I don't think so'.[139]

Wells later stepped back from the extremity of this statement; for example, in his *Experiment in Autobiography* (1934) he allowed that both he and James were 'incompatably [sic] right'.[140] He was correct here, though he did not develop this idea further: the common concern of both Wells and James with movement across multiple perspectives stemmed from a shared interest in inducing a particular state of aesthetic awareness in the reader. The works of both men self-consciously resist passive consumption. 'The rhetoric of spatial relations' features heavily in both writers: the hierarchy implied by much of this spatial rhetoric, including the dimensional analogy, points to a distinction between those 'readers' who can access this higher aesthetic and those who cannot.[141] Mark McGurl observes the contradictory impulses in his analysis of the dimensional analogy in *Flatland*:

> The ideological significance [...] is still an open question: Does it suggest that seen from a higher dimension, Flatlanders achieve a moral equivalence more important than their class differences, that they are equal in the eyes of an all-seeing God? Or does it rather suggest the existence of an ultraexclusive space that might be inhabited by a few godlike persons – a technology of social distinction [...]? The answer [...] is that it is able to imply both of these positions at the same time.[142]

For Hinton, as an egalitarian popularizer of science, this 'pathos of distance' was available to anyone willing to undertake his project.[143] In the case of Wells and Henry James, the matter was more complicated. While Wells had his Over-minds and Samurai (a priestly class of intellectual aristocrats first introduced in *A Modern Utopia*), James had his central consciousness, the character who sees and understands the most and whose level of privileged knowledge is closest to James's own. Both writers used these figures implicitly to raise their readers' level of awareness.

The difference between Wells and Henry James was one of degree, not kind. With his splintering frame technique, Wells wanted to forcibly enact in his readers the kind of perspectival shift that would generate an awareness of – and, by implication – access to, this 'ultraexclusive space'. In James's fiction, particularly his later period, it is the individual central consciousness that accesses the privileged higher perspective, transcending the limited consciousnesses of the other characters; arguably, the perspective of the Jamesian central consciousness most closely resembles James's own from his position in the extratextual dimension. While Wells, particularly in his fiction, attempted to create a four-dimensional reader by disrupting the narrative frame of the text, James dramatized one character's growing awareness of this other dimension. Any critique of the James and Wells debate that attempts to establish hard and fast divisions between the literary

---

139   Wells, *Boon*, 94.
140   Wells, *Experiment*, 2: 493.
141   McGurl, *The Novel Art*, 56.
142   Ibid., 61.
143   Nietzsche, *Beyond Good and Evil*, 192.

aesthetics of either writer will fall short, because, as Sarah B. Daugherty rightly observes, James's 'critical writing [...] is less notable for its syntheses than for its oscillations', and the same can be said of Wells.[144] However, both writers clearly appreciated the others' work enough that, as Wells later put it, they 'bothered' each other.[145]

In his study of Gertrude Stein's writing and its relationship to William James's ideas, Steven Meyer makes the case that 'instead of being modelled on scientific experimentation, her writing turns out to be a form of experimental science itself. [...] She reconfigured science *as* writing and performed scientific experiments *in* writing'.[146] Interestingly, this insight allows us to rethink Wells's splintering frame technique as well. Scientifically trained like Stein, Wells shifted from science teaching and writing to fiction and literary criticism early in his career. His attempts to motivate the reader to self-reflection and social change through the splintering frame technique became a kind of experiment in hyperspace philosophy. Just as in Hinton's *Scientific Romances* and cube exercises, Wells frequently disrupted the narrative flow of his story to call the reader's attention to the world outside the text. The reader is given instructions for a mental or physical task, or is pushed from one discourse into another (see, for example, Hinton's frequent interruptions of 'The Persian King' to explain physical science theory, or Wells's introductory frames for *A Modern Utopia*). In these experiments in hyperspace philosophy, the reader is the subject: just as the plane beings of flatland narratives are often subjected to the epistemological violence of either the sudden intrusion of a hyperbeing or the equally sudden removal into another dimension, the reader of a Wellsian splintered narrative is abruptly brought to awareness of an extratextual dimension.

Though Wells was concerned with exposing the artificiality of conventional narrative structure, his work should not be read anachronistically as poststructuralist or postmodernist.[147] While there is a shared lack of nostalgia for the past in both Wellsian and postmodernist aesthetics, Wells's splintering frame technique, when read through the lens of the fourth dimension, reveals a clear bifurcation between the assumed world of the text and the external 'real' world of the reader. Wells accepted the nineteenth-century scientific epistemological dichotomy between the object and subject. While this may seem a contradictory claim, as Wells was continually drawing attention to his authorial status as a subjective shaper of events in his novels (which continually vexed James), the impact of his narrative intrusions is predicated upon the very assumption of the separation they were created to violate. Thus, in order for his fictions to work as scientific experiments, some fundamental distinction between observer and observed, textual and extratextual, must be maintained. Although desiring to transfigure nineteenth-century narrative conventions, Wells still presented the quest for knowledge as penetrative and invasive.

Perhaps surprisingly (as *The Turn of the Screw* would become a favourite of the New Critics), Henry James's theory of literary art most closely approaches the poststructuralist

---

144 Daugherty, 'James and the Ethics of Control', 64.
145 Wells, *Experiment*, 2: 488.
146 Meyer, *Irresistible Dictation*, xxi, original emphasis.
147 See, for example, Hardy, 'Wells the Poststructuralist'; and Caldwell, 'Time at the End of Its Tether'.

creed that 'there is nothing outside the text'.[148] It was in his final letter to Wells that James made his famous claim:

> So far from that of literature being irrelevant to the literary report upon life, and to its being made as interesting as possible, I regard it as relevant in a degree that leaves everything else behind. It is art that *makes* life, makes interest, makes importance, for our consideration and application of these things and I know of no substitute whatever for the force and beauty of its process.[149]

For James, art creates our entire world; 'reality' is mediated through each individual's process of consciousness. As an author, James's primary vocation was to perform the function of a selective and constructive consciousness of the highest aesthetic refinement possible.

Wells understood the difference between his theory of fiction and James's more clearly than critics usually acknowledge. In the extant fragment of his response to the above letter, Wells wrote:

> I don't clearly understand your concluding phrases – which shews no doubt how completely they define our difference. When you say 'it is art that *makes* life [...]', I can only read sense into it by assuming that you are using 'art' for every conscious human activity. I use the word for a research and attainment that is technical and special.[150]

Wells was correct here; in his letter Henry James used 'art' to indicate the constructive nature of consciousness, something his brother William also accentuated in his radical empiricism. However, Wells – like William – viewed literature as something separate; it was just one particular 'science' of interpreting and representing the world. Thus, Wells was conducting a kind of psychological experiment on his readers; the penetrative aesthetic of the splintering frame is supposed to function as a catalyst for raising the reader to a higher dimension of consciousness. Conversely, Henry James celebrated the unity of the narrative; the result is a hyperreal emphasis on surface where the act of 'reading' – of aesthetic perception – becomes a means of lowering the 'dam' for the Jamesian central character, so that it may momentarily 'spill over' into the extratextual dimension, where James himself resides.

---

148 Derrida, *Of Grammatology*, 163.
149 Edel and Ray, eds, *Henry James and H. G. Wells*, 267, original emphasis.
150 Ibid., 267, original emphasis.

Chapter Six

EXCEEDING 'THE TRAP OF THE REFLEXIVE': HENRY JAMES'S DIMENSIONS OF CONSCIOUSNESS

In the preface to his short story 'The Pupil' (1891), Henry James claimed,

> All I have given [...] is little Morgan's troubled vision of [his family] as reflected in the vision, also troubled enough, of his devoted friend. The manner of the thing may thus illustrate the author's incorrigible taste for gradation and superpositions of effect; his love, when it is a question of a picture, of anything that makes for proportion and perspective, that contributes to a view of *all* the dimensions. Addicted to seeing 'through' – one thing through another, accordingly, and still other things through *that* – he takes, too greedily perhaps, on any errand, as many things as possible by the way. It is after this fashion that he incurs the stigma of labouring uncannily for a certain fulness of truth – truth diffused, distributed and, as it were, atmospheric.[1]

James's explicit statement of interest in refracted perception, on looking *through* multiple lenses of consciousness, is implicitly supported by the complex structure of the sentences here. Writing of Henry James's later style, Hazel Hutchison argues that while his 'late novels are primarily concerned with problems of expression and form, and should be read through the lens of Modernism [...] this position confines James's work to the trap of the reflexive and limits its ability to inform or explain anything outside itself'.[2] At issue here is the problem of how to combine the transcendental with the material: if there is nothing outside of text, is there any way to rise above the surface of textuality, to see *through*? In James's formulation above, 'a view of *all* the dimensions' would perhaps allow 'a certain fulness of truth' to be obtained by the author. Here James was writing about a story from his middle period, where it is specifically the author who is allowed the privilege of viewing through multiple lenses of consciousness; in his later fictions, he attempted to dramatize the experience – through his central characters' developing consciousness – of possessing a view of *all* the dimensions.

In examining James's work through a four-dimensional lens, I am not proposing that James was a hyperspace philosopher. Indeed, there is no evidence that Henry James read Hinton's work. However, he had certainly heard of Hinton's fourth dimension; in addition to his brother's correspondence with Hinton and familiarity with his ideas, several

---

1 H. James, *The Art of the Novel*, 153–54, original emphasis.
2 Hutchison, *Seeing and Believing*, 4.

members of Henry James's circle of literary friends at the turn of the century were discussing and writing about Hinton's ideas. In addition to H. G. Wells, Joseph Conrad and Ford Madox Ford collaborated on a novel about four-dimensional socialist revolutionaries in *The Inheritors* (1901), and Henry James reviewed *The Martian* (1897), a novel by another friend, George du Maurier, which makes use of the fourth dimension as well. As Joan Richardson observes, the topic of higher dimensions was 'part of the cultural conversation conducted in the reviews and journals to which Henry James was a regular contributor' and 'the news of these significant challenges to inherited notions of space, time, and perception was certainly not lost on James'.[3] Therefore, if we seek a productive way of considering James's later work as participating in the developing concerns of early modernism while avoiding the trap of a reductive focus on the labyrinthine reflexivity of his style, it is productive to look at it through the lens of the fourth dimension and particularly the dimensional analogy. Indeed, such an analysis reveals a number of parallels between the dimensional analogy and the hierarchies of perception in James's fiction.

My focus in this chapter is on the texts from what could be described as James's transitional phase, between his so-called middle and later periods. It was during the late 1890s and early 1900s that James was most interested in Wells's writing; as their correspondence reflects, it was during this time that James was carefully reading and mentally 'rewriting' Wells's work. Henry James was unimpressed with Wells's attempts to disrupt the unity of the narrative frame via his developing splintering frame technique. In his first letter after reading *Boon*, James reiterated his belief that 'the fine thing about the fictional form to me is that it opens such widely different windows of attention; but that is just why I like the window so to frame the play and the process!'.[4] James's ideal sacrosanct narrative frame became increasingly permeable as the perceiving consciousness of his narratives became more and more expansive.

In James's later writing, we increasingly find moments of revelation where a character approaches a view of *all* the dimensions, like the author. The boundary between what is contained within the text (James's central character) and the extra-textual author (James himself) must be maintained to avoid a total collapse into unmediated authorial intrusion into the text, either in the Trollopian aside[5] or the 'first person', 'that accurst autobiographic form which puts a premium on the loose, the improvised, the cheap and the easy'.[6] Grappling with this problem when composing *The Ambassadors*, James argued that allowing Lambert Strether the first-person perspective was not an option, as the first person, employed in this way, 'is addressed by the author directly to ourselves, his possible readers'.[7]

However, as the central characters of James's fictions labour 'uncannily for a certain fulness of truth', at times it seems as if the narrative frame will not hold. These are moments of intense perception – to borrow William James's words – 'odd lowerings

---

3 Richardson, *A Natural History*, 153–54.
4 Edel and Ray, eds, *Henry James and H. G. Wells*, 263.
5 Matthiessen, ed., *The James Family*, 355.
6 Edel and Ray, eds, *Henry James and H. G. Wells*, 128.
7 H. James, *The Art of the Novel*, 321.

of the brain's threshold', where the central characters themselves appear to channel extra-textual awareness into the narrative frame.⁸ Read through the lens of the dimensional analogy, in these moments of intense perception, Henry James's central character momentarily experiences a sense of an extra-textual dimension, which, when compared with Wells's splintering frame technique, represents a dramatization of the perspectival shift that Wells attempts to enact on the reader. The difference here is comparable to the two characters who become aware of the action of Hinton's Persian king: one (the prince) is the recipient of the king's revelation, while the other (the student) must painstakingly work it out for himself.

In the present chapter I explore the moments in fictions from James's transitional period where his central characters work 'it' out, as Fleda Vetch does so 'ingeniously' in *The Spoils of Poynton*.⁹ While I touch on examples from a number of James's texts from this period, I look most closely at three texts from the turn of the century, *The Spoils of Poynton* (1896–1897), 'The Great Good Place' (1900) and a later text, 'The Jolly Corner' (1910), which James began developing in 1895. Before I begin my analyses of these particular fictions, however, I look more closely at the shifting formal quality of James's fiction from this period. Again, it is useful to recall Wells's four-dimensional splintering frame aesthetic, as James was attempting to find a more elegant solution to the problem of increasing perspectivism through the outward-spiralling consciousness of his central character.

The Spiralling Consciousness in Henry James

Writing of *The Awkward Age* (1898–1899), Daniel Mark Fogel has drawn attention to the underlying pattern of James's most structurally schematic novel:

> The diagram James drew of the ten books through which the intrigue is played out – a drawing, as he relates, of 'small rounds' in 'the neat figure of a circle' – implies the pattern of the spiral dialectic, rounding back to its point of origin.¹⁰

The Awkward Age – which begins with its first book, 'Lady Julia' – ends with a book named after Lady Julia's granddaughter and modern replica, Nanda. Here we have a circling back to origins, but not without alteration. Nanda is *not* Lady Julia; her circumstances do not allow for her to be an *exact* replica of her grandmother. Rather, Nanda is an expression of her grandmother's spirit and temperament, as exposed to and influenced by the modern mores of her mother's social circle in London. The differences between Nanda and Lady Julia are therefore a result of their divergent experiences.

There is repetition with variation here, reminiscent of Emerson's 'system of concentric circles', undergoing occasional 'slight dislocations, which apprize [*sic*] us that this surface on which we now stand is not fixed, but sliding'.¹¹ However, the pattern in James's

---

8 W. James, *Writings, 1878–1899*, 1119.
9 H. James, *The Spoils*, 196.
10 Fogel, *Henry James and the Structure of the Romantic Imagination*, 15.
11 Emerson, *Essays and Lectures*, 409.

fiction is more aptly described as that of a spiral rather than a circle.[12] This spiral pattern of subtle expansion of human consciousness in James can be read as an updated version of Emerson's expanding circles, a connection Jonathan Levin makes in his description of the 'poetics of transition' in both writers: 'the Emersonian self is [...] a circle always expanding into broader circles by means of ceaseless crossings and transitions'.[13] The figure of the spiral emphasizes the 'sliding' nature of the transitions between levels of consciousness, rather than the metaphoric 'jumps' required between Emerson's set of concentric circles.

A contemporaneous description of this spiral pattern can be found in Wells: in an 1894 essay titled 'The Cyclic Delusion', Wells claimed that apparent cycles identified in contemporary biological, mathematical and astronomical theories 'seem cyclic only through the limitation of our observation'.[14] Wells went on to argue that natural cycles repeat, but always with variation. Species change over time, accumulating various traits until they are entirely distinct from their ancestors. This movement, of 'circular repetition and change, a recurrence with variation', as Scheick describes it, 'graphically [...] may be depicted as a spiral'.[15] The pattern of the spiral, Scheick continues, also accurately represents the impulse behind Wells's splintering frame technique. He explains that,

> as the intrinsically four-dimensional human mind increasingly clarifies, it expands the circumference of the preceding framework circumscribing that mind's thought; this enlarged ring, as it were, becomes the new framework, which in turn is to be splintered outwardly and so on ad infinitum.[16]

Wells's experiments in splintering the frame of the text are, in effect, attempts to push his readers to attain a higher level of consciousness, or 'a view of *all* the dimensions'. While in his later texts Wells was interested in the utopian possibilities of achieving such a view, in his earlier scientific romances he tended to focus on the more ambiguous effects of increased knowledge. Likewise, the ending of *The Awkward Age* implies that Nanda is worse off for her expanded consciousness of the morally bankrupt society of her mother; Nanda's 'modernity' lies in the fact that she is *not* sheltered from the adulterous intrigues of her elders.

The spiral pattern of expanding consciousness can be found in other James novels. As J. A. Ward observes, in much of James's fiction, 'the structure derives not from the external action, but from the developing awareness of the central consciousness'.[17] It is easy to imagine the structure of this development as a spiral, because, Ward argues, the 'action' in James is often progressively repetitive: 'The rhythm of *The Ambassadors* is both repetitive and progressive. But the outward action is almost wholly repetitive; the progressive is limited to the growth of Strether's knowledge'.[18] The only change in the recurring cycle

---

12 See also Ward, '*The Ambassadors*', 366 and 368.
13 Levin, *The Poetics of Transition*, 30.
14 Wells, *H. G. Wells: Early Writings*, 112.
15 Scheick, *The Splintering Frame*, 32.
16 Ibid., 33.
17 Ward, *The Search*, 42. See also Levin, *The Poetics of Transition*, 118.
18 Ward, *The Search*, 47.

of dinners, parties and conversations in which Strether participates throughout the novel comes from his constant reassessment of his understanding of Chad's circumstances. With each revision, Strether's view of the situation in Paris grows in a way that is analogous to the change in perspective that occurs with the shift to a higher dimension, or an expanded frame. James even staged the scene of Strether's key realization of Chad and Madame de Vionnet's affair as a violation of his perceptual frame, a kind of 'Through a Window' moment where Strether's aestheticized pastoral fantasy is disrupted.

On his train ride out of Paris into the French countryside where he intends to experience 'that French ruralism [...] into which he had only looked through the little oblong window of the picture frame', Strether literally steps into his imagined model of rural France, one influenced by a Lambinet painting he had seen several years earlier at a Boston art dealer's. Strether recalls the painting as 'the picture he *would* have bought – the production that had made him for the moment overstep the modesty of nature'. The purchase had been beyond his meagre means: 'a price he had never felt so poor as on having to recognise, all the same, as beyond a dream of possibility'.[19] His recollection of the painting as a part of his past, of the New England puritanical values he represents as an 'ambassador' in Paris, of 'the maroon-coloured, sky-lighted inner shrine of Tremont Street [...] the dusty day in Boston' when he saw the painting with 'the ridiculous price, the poplars, the willows, the rushes, the river', allows Strether not only to step into the pastoral fantasy of the French countryside now present to him, but into his personal past, which is transposed onto the present:

> The train pulled up just at the right spot, and he found himself getting out as securely as if to keep an appointment. It will be felt of him that he could amuse himself, at his age, with very small things if it be again noted that his appointment was only with a superseded Boston fashion. He hadn't gone far without the quick confidence that it would be quite sufficiently kept. The oblong gilt frame disposed its enclosing lines; the poplars and willows, the reeds and the river [...] fell into composition, full of felicity, within them; the sky was silver and turquoise and varnish; [...] it was all there, in short – it was what he wanted: it was Tremont Street, it was France, it was Lambinet. Moreover he was freely walking about in it.[20]

Strether is ambulating through his memory of the Lambinet painting, 'boring so deep into his impression and his idleness that he might have fairly gotten through them again and reached the maroon-coloured wall' of the art dealer's shop on Tremont Street in Boston. He is so thoroughly in possession of the impression, however, that he 'not once overstep[s] the oblong gilt frame. The frame had drawn itself out for him, as much as you please'.[21] However, again the 'price' of this experience will interfere with his total possession of it.

His moment of 'revelation' occurs at a riverside café, the Cheval Blanc, where he is given 'a sharper arrest' when he sees 'just exactly the right thing', the figures that 'had

---

19 H. James, *The Ambassadors*, 380, original emphasis.
20 Ibid., 381.
21 Ibid, 385.

been wanted in the picture': a man and a woman in a boat on the river, approaching the café.[22] The figures are Chad and Madame de Vionnet, who have escaped Paris for an illicit tryst and they literally invade the frame of Strether's pastoral scene, as Alexander Gelley observes: 'What Strether sees when he recognizes the guilty couple dispels the sphere of "fancy" or "art" and brings into play a moral dimension'.[23] Here is a refined dramatization of the violation of a perceptual frame, what Wells literalized in 'Through a Window'. Strether is abruptly reminded of the world outside of the 'text'; the source of his tension in Paris literally floats into Strether's imagined pastoral painting. Just as Wells wrote of the importance of 'scepticism of the instrument', in the Lambinet scene, the incongruence of Strether's 'perceptive faculties' (signified by his entrance into the frame of a picture he associates with Boston) and the illicit elements of Chad's 'virtuous attachment' are brought sharply to his attention. After this revelation, Strether is raised to 'a view of *all* the dimensions'.

Writing of Strether's adventure of consciousness throughout *The Ambassadors*, Nicola Bradbury observes,

> Strether's misunderstandings arise from his attempt to 'place' the Parisian situation in a New England framework of moral judgement, and to interpret the behaviour of those in Paris according to the patterns of New England custom and prejudice. [...] In the Lambinet episode he commits the opposite fault, in drawing a fanciful frame around a dramatic scene, and thus mistaking the 'formal context' of the phenomena he witnesses. It is a recognition which marks Strether's full consciousness: not new experience, but the placing of his perceptions in perspective, within the right frame.[24]

By entering the Lambinet painting of his memory, Strether is able to see 'through one thing' – his New England consciousness – 'to another'. His abrupt reminder of the 'moral dimension' of his trip to France does not lead him to condemn the affair between Chad and Madame de Vionnet, as he was prepared to do when he first arrived in Paris. Bradbury aptly identifies 'the paradox of his "double consciousness" [that] places Strether outside the rigour of the mutually exclusive categories of [New England] and Paris alike' as 'the first glimpse of a transcendent consciousness which will free him from both worlds'. Strether transcends the limitations of his own fancies here, and 'though he remains a romancer', his 'vision has come to equal that of the narrator', or, indeed, James himself.[25] Paradoxically, Strether's transcendence of his limitations as a represented, literary character is obtained by entering 'into' and passing 'through' his aesthetic vision.

If we examine the Lambinet scene through the lens of the dimensional analogy, the implications of James's aesthetic are brought into focus. Hazel Hutchison observes that 'at the height of his powers of perception, [Strether] can recognize his situation only through the agency of a long absent picture, which revises and reinvents itself as

---

22 Ibid., 388.
23 Gelley, 'The Represented World', 421.
24 Bradbury, *Henry James: The Later Novels*, 40.
25 Ibid., 41 and 62.

he engages with it'.²⁶ While Hutchison makes an excellent point here about Strether's engagement, I worry that her attribution of agency to the picture (its ability to revise and reinvent *itself*) de-emphasizes the fact that it is Strether who is sustaining this representation, actively revising it and reinventing it as he ambulates 'through' it. What is important here is that James's dramatization of aesthetic engagement affirms a belief 'that art was something to be stepped into and completed by its beholders'.²⁷ It is through 'stepping' into art that the Jamesian central character is able to see more than all the other characters; this is a view of all the dimensions only James himself shares. In this sense, the Jamesian central character enters a literary fourth dimension, a fictive space of heightened consciousness beyond the world of the text that contains them. At times Strether appears to be, as Richard Salmon notes, 'simultaneously and self-consciously inside and outside the frame of his own representation'.²⁸ The ability to view a situation from an authorial perspective is, in many ways, the pinnacle of achievement for the Jamesian central character, who, according to Timothy Lustig, 'in its quest for subjective freedom [...] is at some level aspiring to the position occupied by James the writer'.²⁹

The Spoils of Poynton (1896–1897)

One central character who achieves this subjective freedom – at least temporarily – is Fleda Vetch. In his preface to *The Spoils of Poynton*, James described Fleda's perception as 'almost' demonic.³⁰ As I observed in Chapter Two, the demonic is a useful imaginative trope; as an uncanny figure of ambulation, it is able to traverse boundaries that would otherwise remain impenetrable. With her 'almost demonically' heightened consciousness, Fleda is able to cross the boundary between the represented, fictional space in *The Spoils* and James's extra-textual perspective, achieving a subjective freedom similar to that of the student in Hinton's 'The Persian King'. The other characters with whom Fleda shares the text are not even aware of the possibility of this kind of autonomy, as James notes in the preface:

> Thus we get perhaps a vivid enough little example, in the concrete, of the general truth, for the spectator of life, that the fixed constituents of almost any reproducible action are the fools who minister, at a particular crisis, to the intensity of the free spirit engaged with them. [...] The free spirit, always much tormented, and by no means always triumphant, is heroic,

---

26 Hutchison, *Seeing and Believing*, 94.
27 Siraganian, 'Out of Air', 661. Siraganian examines Lewis's satire of this aesthetic of engagement by having two characters literally step into a painting in his 1928 novel, *The Childermass*. A year previous to this novel, Lewis published *Time and Western Man*, a book-length critique of what he called 'time-philosophy', or the 'time-cult', and its expression as 'art that needs a spectator's experience to be whole' in post-Impressionist art and literature (661). Lewis cited Hinton as a forerunner of this 'time-philosophy'; see Lewis, *Time and Western Man*, 198.
28 Salmon, *Henry James and the Culture of Publicity*, 176.
29 Lustig, *Henry James and the Ghostly*, 63.
30 H. James, *The Art of the Novel*, 129.

ironic, pathetic or whatever, and, as exemplified in the record of Fleda Vetch, for instance, 'successful', only through having remained free.[31]

Unlike these 'fixed constituents', Fleda has the ability to roam between multiple viewpoints: at different times she empathizes with Owen, Mrs Gereth, Mona, Poynton, the deceased maiden aunt at Ricks and even – as I will argue – James himself. To borrow the language of Forster, she is 'round' while the others are flat; she possesses an extra dimension.

This ability to both see and feel in multiple ways allows Fleda to choose a course of action that preserves her autonomy, as Millicent Bell highlights:

> She is sufficiently imaginative to be capable of a variety of choices; she is not determined and predictable like Mrs. Gereth with her single obsession. [...] The 'things,' though she admires them, do not govern her behavior; she has no desire to possess them [...]. Neither does her love for Owen overcome her reluctance to own him; she renounces the desire to possess him as though he were a thing to be wrested from someone else.[32]

Fleda is 'free' because she has transcended the strictly possessive relations in which the other characters are entangled: the other women in *The Spoils* view each other and the amiable but dull-witted Owen as either antagonists or pawns that must be moulded to one's own will.[33] If 'possession threatens the integrity of the conscious self',[34] Fleda maintains her integrity by refusing to succumb to her desire to 'possess' either Owen or Poynton. Her self-possession is thus of the same kind as the student's in 'The Persian King'; he is able to detach his own will from the king's manipulative dynamic system of pleasure and pain. Hinton distinguished such characters as 'true personalities conscious of being true selves, the oneness of all of them lying in the king, but each spontaneous in himself and absolute will, not to be merged in any other'.[35]

Writing of 'The Persian King', Bruce Clarke illustrates the application of the dimensional analogy to a hierarchical understanding of degrees of consciousness: 'the relation of the valley subjects to the king embodies the susceptibility to manipulation, the puppet-like dependency, that would exist in the relation of the three-dimensional being to a hyper-being'.[36] Making a direct comparison between Fleda's 'quasi-authorial intelligence' and the dimensional analogy, Mark McGurl looks to *Flatland*, noting that 'similarly did Abbot's A. Square, rising above his two-dimensional world and seeing it whole, experience the comparative stupidity of those living below'.[37] In one sense, as McGurl observes, Fleda – and all literary characters – are literally Flatlanders like A. Square because they are all confined to the plane surface of the page of the text. It is the combined effort of the author and the reader that 'lifts' them into a 'fleshed-out' – albeit imaginary – 'three-dimensional'

---

31 Ibid., 129–30.
32 M. Bell, *Meaning*, 214–15.
33 See also Mitchell, 'To Suffer'.
34 Hutchison, *Seeing and Believing*, 36.
35 Hinton, 'The Persian King', *Scientific Romances*, 127–28.
36 Clarke, *Energy Forms*, 116.
37 McGurl, *The Novel Art*, 75.

existence. Both Hinton's student and A. Square are conscious of their status as lower-dimensional beings, and it is through this awareness that they establish a certain amount of subjective autonomy. A large part of Fleda's freedom stems from her awareness of something that lies outside of herself, the other characters and even the 'spiritual' aura of Ricks. Read through the dimensional analogy, this awareness can be understood as a connection between Fleda and the author, James. Such moments of connection in Henry James's fiction are often figured – like the supernatural experiences of the extraordinary individuals that his brother William researched – as ghostly encounters.

While critics do not usually number *The Spoils* among James's supernatural tales, there is an element of the ghostly that permeates this text. The 'spoils' themselves – the priceless antique objects that constitute the 'bone of contention' in the novel – provide this aura by being 'conspicuously absent' from the text, as Bill Brown argues:

> Despite the novel's eventual reference to two specific objects – a 'great Italian cabinet' in the red saloon and the exquisite Maltese cross [...] Poynton, above all, is awash in overarching characterization [...]. James's mise-en-scène at Poynton is a matter of aura, not artefacts.[38]

Like Hinton's king, the objects of Poynton remain un-representable; collectively, they are an 'absent presence' recognized by only a select few. This presence provides the invisible locus of James's drama. In his remarks in the preface, James wrote that, for him, Fleda acts as a stand-in for the invisible things of Poynton: 'the real centre [...] would have been the Things, always the splendid Things'.[39] Fleda, then, is not only the mediator between the Gereths, but also between 'the splendid Things' in the author's imagination, and the reader. 'The splendid things', as Brown observes, indicate 'an absent centrality'.[40]

In the preface, James claimed that the limitations of space within the original periodical publication of the novel prevented him from fully describing the spoils. Market issues aside, however, the fact that the objects of Poynton remain visually absent supports Tzvetan Todorov's observation that at the heart of the Jamesian fiction lies 'an absent and superpowerful force' that drives the whole narrative.[41] James was unable or unwilling to describe this 'absent and superpowerful force' directly to his readers; rather, he illuminated it through the eyes of one of his characters. The role of this central consciousness is to gradually reveal the multiple aspects of the situation depicted within the text. This is the role Strether plays in *The Ambassadors* and it is also Fleda's role in *The Spoils of Poynton*. The methodology of refracting the narrative through Fleda's consciousness can easily be described in terms of William James's pragmatic psychology, as Merle A. Williams has observed: like his brother William's model of the ambulating consciousness, *The Spoils of Poynton* is 'a work whose preliminary assumptions are repeatedly revised, interrogated, and redirected in light of fresh evidence'.[42]

---

38 Brown, 'A Thing about Things', 226.
39 H. James, *The Art of the Novel*, 126.
40 Brown, 'A Thing about Things', 230.
41 Todorov, *The Poetics*, 145.
42 M. Williams, *Henry James and the Philosophical Novel*, 164.

This process of reasoning can be imagined as an expanding spiral of consciousness, or as Hinton's 'Arabic method of description' in 'The Persian King', which deployed numerical notation as a metaphor for the means used by the student to perceive the fourth dimension. This pattern of assimilation and adjustment can be applied to many of James's central characters, including Fleda who learns to appreciate the absent presence of the spoils and the author himself. It is telling that James did not describe the spoils physically, or even name them. Fleda's sensitivity to this invisible centre is a sign of her empathy with James the author: 'Fleda's ingratiating stroke for importance […] had been that she would understand; and […] the progress and march of my tale became and remained that of her understanding'.[43] Thus, the true plot of this novel appears to be a charting of Fleda's 'growing consciousness of the whole, or something ominously like it'.[44]

How did James figure this 'consciousness of the whole'? Fleda, appreciator of invisible things, is also able to 'read' the character of the maiden aunt, which lingers in the objects and spaces of the house at Ricks. The maiden aunt never appears in the novel; she has been dead for years, and little information is given about her. Early in the novel, on her first visit to Ricks, Fleda declares the house 'charming', but still finds it necessary to avert her eyes from the tacky ornamentation in the garden and the house's interior.[45] Upon her return to Ricks at the end of the novel, Fleda's response is more dramatic: 'The effect […] arrested her on the threshold: she stood there stupefied and delighted at the magic of a passion of which such a picture represented the low-water mark'.[46]

Fleda's consciousness has developed over the course of the novel to the point at which she is able to read James's characterization of the maiden aunt – 'the magic' of the 'passion', which had been her life – in the 'four sticks' of furniture she has left behind.[47] However, there is something else, something beyond the ghost of the maiden aunt, which permeates and contributes to the atmosphere at Ricks. Tellingly, Fleda describes this presence as an absence, leaving

> 'the impression somehow of something dreamed and missed, something reduced, relinquished, resigned: the poetry, as it were, of something sensibly *gone*'. Fleda ingeniously worked it out. 'Ah, there's something here that will never be in the inventory!'[48]

The text of the inventory of the maiden aunt's belongings at Ricks and the text of the novel in which they all reside, are both superseded here. There is an analogy between the 'something' that Fleda recognizes as lying outside of the text of the inventory of the 'four sticks', and the presence that exists outside of her own text, which – as James notes in the preface – is also reduced to four 'main agents'.[49]

---

43 H. James, *The Art of the Novel*, 128.
44 Ibid., 128.
45 H. James, *The Spoils*, 59.
46 Ibid., 195.
47 Ibid, 196.
48 Ibid., 196, original emphasis.
49 H. James, *The Art of the Novel*, 128.

When asked by Mrs Gereth to give this 'something a name', Fleda responds:

'I can give it a dozen. It's a kind of fourth dimension. It's a presence, a perfume, a touch. It's a soul, a story, a life. There's ever so much more here than you or I. We're in fact just three!'
'Oh if you count the ghosts –!'
'Of course I count the ghosts, confound you!'[50]

The third member of the party at Ricks, it is implied, is the ghost of the maiden aunt. Fleda is able not only to detect this ghost, but also to 'ingeniously and triumphantly' work 'it' out: the sense of 'something' over and beyond the ghost. This 'something', as Lustig notes, can be understood as the authorial presence: another sort of 'ghostly encounter [that] represents a particularly intense adventure of consciousness, an access of liberated and disencumbered experience, and one could argue that it brings the ghost-seer extremely close to James himself'.[51]

Fleda senses a connection between her story and the story of another being haunting the text. In her attempts to describe this connection, she struggles to find a precise vocabulary. She linguistically ambulates around it, substituting a dozen names, and it is appropriate that one of the labels she chooses is 'a kind of fourth dimension'. Fleda's realization that 'there's ever so much more here than you and I', even beyond the ghostly presence of the maiden aunt – whom she includes in the 'you and I' – can be read alongside higher-dimensional parables such as those of Hinton and Wells. There is an implied hierarchy of relations here, between Fleda and the other characters in the novel, and between Fleda and James. Fleda is able to understand and appreciate the invisible centre of the text – the spoils – because her understanding is closest to that of James's. Like Hinton's student, Fleda is also aware of an invisible, conscious agent above and beyond herself. While James did not puncture the fictional world of this novel in the way that Wells splintered the frame of his own narratives, he managed to escape the 'trap' of pure reflexivity by projecting his presence into the text through Fleda's recognition of the 'soul' or 'life' beyond herself, Mrs Gereth and the ghosts of Ricks.

'The Great Good Place' (1900)

Turning to a different work from this transitional moment between his middle and late periods, we observe James taking another protagonist on an 'intense adventure of consciousness', this time outside of the real, 'lived' space of the text into a 'place' that seems to exist outside of representational time and space altogether. This place is perhaps the most heavily infused with authorial presence within James's œuvre. James's characters in this story also give this place multiple names, but for the main protagonist to whom

---

50 H. James, *The Spoils*, 196.
51 Lustig, *Henry James and the Ghostly*, 63.

James grants access to this 'liberated and disencumbered experience', it is known simply as 'The Great Good Place'. I will consider the space of 'The Great Good Place' as legible as a 'kind of fourth dimension' – a literary fourth dimension – which refigures this transcendental, abstract space as the place of the creative imagination.

Years after James's death, Ford Madox Ford recalled a conversation he had with James, concerning James's story, 'The Great Good Place'. James and Ford had been out walking one day, Ford recalls, when James began:

> 'There are subjects one thinks of treating all one's life. […] And one says they are not for one. And one says one must not treat them […] all one's life. All one's life. […] And then suddenly […] one does. […] *Voilà!*' [James] had been speaking with almost painful agitation. He added much more calmly: 'One has yielded to temptation. One is to that extent dishonoured. One must make the best of it.'[52]

According to Ford, in writing 'The Great Good Place', James

> considered he had overstepped the bounds of what he considered proper to treat – in the way of his sort of mysticism. […] For there were whole regions of his character that he never exploited in literature, and it would be the greatest mistake to forget that the strongest note in that character was a mysticism different altogether from that of the great Catholic mystics. It resembled rather a perception of a sort of fourth dimensional penetration of the material world by strata of the supernatural.[53]

In 'The Great Good Place', the space of the Place functions as a kind of 'non-representational space' where the central character, the writer George Dane, is allowed to retreat and regain his sense of self – his own creative agency as an author – through communion with James's authorial consciousness. Unlike Fleda, Dane's world is not temporarily infused with 'a sort of fourth dimensional penetration' from the extra-textual realm of the author; rather, Dane is removed *to* this place. Once there, Dane experiences several 'odd moments' of encounter with some unseen agent:

> [Dane] analysed, however, but in a desultory way and with a positive delight in the residuum of mystery that made for the great agent in the background the innermost shrine of the idol of a temple; there were odd moments for it, mild meditations when, in the broad cloister of peace or some garden-nook where the air was light, a special glimpse of beauty or reminder of felicity seemed, in passing, to hover and linger.[54]

Dane's ghostly encounters with this agent are certainly more frequent than Fleda's, and they are seemingly uninitiated by himself. Unlike Fleda, whom James described as 'planting herself centrally' in the narrative almost as if she possessed an agency beyond that of

---

52 Ford, *Mightier than the Sword*, 25.
53 Ibid., 25–26.
54 H. James, *Novels and Tales*, 16: 251.

the author, Dane has moved to a place beyond the text where the authorial agent hovers and lingers at will.[55]

At the beginning of the story Dane feels overwhelmed by his worldly success as a writer; his professional and social relations – letters from friends and enemies, journals and books to be reviewed, breakfast and lunch dates – stifle his enjoyment of life, and with it, his creativity. Dane dreams of escaping these relations, and salvation comes when a younger struggling writer visits him. At this crisis point in his life, wishing to 'think, to cease, [...] to do the thing itself', Dane jumps at the younger man's offer to take on the burden of his work while he escapes to the Place.[56] It is this interaction with the younger writer that transports Dane to the Place, and under this 'spell', Dane comes to consciousness in the Place.[57] He is a guest in the boarding house of this Place, where James seems to constantly intrude upon Dane's 'mild meditations', to 'hover and linger' at will. There is no violence to these intrusions, however. Unlike Strether's shock as his Lambinet scene is invaded or Fleda, who is 'arrested' at the threshold of Ricks, within the chimerical borderland space of the Place there is a perpetual sound of 'slow footsteps', of 'a quiet presence' passing somewhere in the background.[58] It seems that here James is the 'ghost' that haunts the Place, and, paradoxically, his presence within the text makes this story somehow more 'realistic' than his more traditionally mimetic fiction. James, infiltrating the space of his own text, brings with him realism of a different order.

James was uncompromising about the relation that the reader of 'The Great Good Place' must have to this text:

> There remains 'The Great Good Place' (1900) – to the spirit of which, however, it strikes me, any gloss or comment would be a tactless challenge. It embodies a calculated effect, and to plunge into it, I find, even for a beguiled glance – a course I indeed recommend – is to have left all else outside. There then my indications must wait.[59]

Ford's observation that James was reticent about 'The Great Good Place' is reflected here in the preface to the New York edition. In this brief, cryptic remark on the story, James's assertion that 'any gloss or comment' on its spirit 'would be a tactless challenge' parallels Ford's reconstruction of his comments on the piece. James also remarked that he intended to embody a 'calculated effect' in 'The Great Good Place', but he did not offer us any clues here as to his intentions. However, if we consider his comments in the preface to *The Spoils of Poynton* in light of Ford's recollections about James's reticence concerning the subject of 'The Great Good Place', we can venture a possible conclusion. Writing in reference to Fleda, James remarked:

> One is confronted obviously thus with the question of the importances; with that in particular, no doubt, of the weight of intelligent consciousness, consciousness of the whole,

---

55 See H. James's description of Fleda in *The Art of the Novel*, 127.
56 H. James, *Novels and Tales*, 16: 246.
57 Ibid., 247.
58 Ibid., 236 and 241.
59 H. James, *The Art of the Novel*, 237.

or of something ominously like it, *that one may decently permit a represented figure to appear to throw.*⁶⁰

In *The Spoils of Poynton*, Fleda is given a hint of 'a presence, a perfume, a touch', which she describes as 'a kind of fourth dimension', and could be read as James's authorial presence. While this 'supernatural' or extra-textual presence only briefly penetrates the 'material' world of James's story in *The Spoils of Poynton*, in 'The Great Good Place' James took Dane entirely out of the 'material' world and placed him into constant contact with this presence. Perhaps – in allowing the reader to experience this 'Place' through the consciousness of Dane – James felt he had overstepped the bounds of what a fictional character should be allowed to reveal.

The beginning of the second chapter of 'The Great Good Place' informs us that Dane is now in the Place, but as William McMurray highlights, 'how much time has lapsed and where the Place is, however, neither Dane nor the reader learns, for time and place seem not to exist'.⁶¹ Indeed, Dane appears to have stepped outside of regular time and space upon his entry into the Place. He dreamily muses on the constant sound of bells in the Place:

> How could they be so far and yet so audible? How could they be so near and yet so faint? How above all could they [...] be, to *time* things, so frequent? The very essence of the bliss of Dane's whole change had been precisely that there was nothing now to time.⁶²

In the Place, Dane seems to have transcended the standard, lived time of 'reality'; however, he remains aware of its passing somewhere off in a removed and un-locatable space. In an exchange with another inhabitant of the Place, Dane attempts to ascertain their geographical location:

> [Dane] asked with the first articulation as yet of his most elementary wonder:
> 'Where is it?'
> 'I should n't be surprised if it were much nearer than one ever suspected.'
> 'Nearer "town," do you mean?'
> 'Nearer everything – nearer every one.'
> George Dane thought. 'Would it be somewhere [...] down in Surrey?'⁶³

Dane, a new arrival to the Place, at first attempts to frame it within the terms of his experience in the outside world. He clearly displays the narrowness of his current view when he substitutes London ('town') for the universe, and attempts to locate the Place as a bourgeois retreat somewhere in the home counties. I am inclined to agree with W. H.

---

60 Ibid., 128, emphasis added.
61 McMurray, 'Reality in James's "The Great Good Place"', 83.
62 H. James, *Novels and Tales*, 16: 236, original emphasis.
63 Ibid., 238.

Auden's assessment that, in 'The Great Good Place', James 'is not describing some social Utopia, but a spiritual state that is achievable by the individual'.[64] Unlike Wells in his later socialist fantasies, James dramatized the individual's experience of the Place as a state of being where time and space are reconfigured; it is allegorized as a place, a sort of club that is 'nowhere' and yet near everything, and is only accessible to a privileged, creative few.

It is possible for the 'location' of the Place to be 'nearer everything – nearer everyone' because it does not inhabit a typically material space. Like the transcendental fourth dimension of hyperspace philosophy, the Place might lie right next to three-dimensional everyday reality but remain unnoticed because in order to discover it one would have to 'learn' to look for it in a way that is currently unimaginable to the average individual. When the time arrives to leave the Place, Dane worries that he will never be able to find his way back again: 'Was this a threshold perhaps, after all, that could be crossed one way?'.[65] Here one is reminded of A. Square of *Flatland*, who attempts to remember how to access the third dimension of space by desperately repeating 'Upward, not Northward'.[66]

From Dane's side of the threshold within the Place, he is able to look outward onto a world that is unaware of his location. '"The thing was to find out!" [...] "And when I think," said Dane, "of all the people who have n't and who never will!" He sighed over these unfortunates'.[67] Dane is in the position of a spectator here. When he first finds himself in the Place, he observes:

> This was the part where the great cloister, enclosed externally on three sides and probably the largest lightest fairest effect, to his charmed sense, that human hands could ever have expressed in dimensions of length and breadth, opened to the south its splendid fourth quarter, turned out the great view an outer gallery that combined with the rest of the portico to form a high dry loggia.[68]

One side of the cloister opens out to form a high loggia, which provides a 'great view' over the countryside. Dane's position is thus one of an elevated and detached observer whose perspective on the world is from a place – it is described as 'the scene of his new consciousness' – that is outside of conventional time and space.[69] During a conversation with a 'Brother' in the Place, Dane and his companion sit together, observing 'the vague movements of the monster – madness, surrender, collapse – they had escaped' in the outside world, from their bench, which 'was like a box at the opera'.[70] They are spectators, rather than characters on the stage of representable 'reality'.

---

64 Auden, *The Dyer's Hand*, 322.
65 H. James, *Novels and Tales*, 16: 259.
66 Abbott, *Flatland*, 191–97.
67 H. James, *Novels and Tales*, 16: 237.
68 Ibid., 235.
69 Ibid., 234.
70 Ibid., 239.

Like Strether, whom Salmon suggests is 'perhaps the fascinated spectator of his own display, gazing at the very scene that awaits him',[71] Dane's position here brings a number of other Jamesian spectators to mind. For example, we can recall Maggie's position on the terrace near the end of *The Golden Bowl*, where she finally understands the complex web of adultery between her husband, father and stepmother/best friend. Looking to even earlier Jamesian fiction, McGurl writes of the first meeting between Hyacinth and the Princess in the theatre box in *The Princess Casamassima*: 'Here it is as though the fourth wall has fallen, actors and audience, bookbinders and princesses, [are] now dwelling in the same order of space'.[72] The fourth wall of Dane's cloister is literally missing, but the effect is the opposite from Hyacinth's entry into the theatre box, which brings him into the world of fiction (he is in such close proximity to the stage that he cannot speak without disturbing the performance) and into a relationship with a princess who is, as McGurl reminds us, the only one of James's major characters to stray from the pages of one novel (*Roderick Hudson*) to another.

While the resurrection of characters within a novelist's œuvre is a common device of high-Victorian realism, the language James used to describe the Princess's wandering between the two novels is suggestive of something beyond the ken of typical mimetic fiction. Quoting James's prefatory discussion of the *The Princess Casamassima*, McGurl notes that the revival of the Princess functions

> as a penetration of fictive space, where by some 'obscure law' certain 'of a novelist's characters, more or less honourably buried, revive for him by a force or a whim of their own and "walk" round his house of art like haunting ghosts, […] pressing their pale faces, in the outer dark, to lighted windows'. Hyacinth at the window of his sweetshop, the ghostly Princess at the lighted window at the 'house of art', befriend each other in the uncanny theater of fiction.[73]

In 'The Great Good Place' we see another of James's characters cut loose from the tethers of the text: the dissolution of the 'fourth wall' in Dane's world allows him to step into the backstage area of the uncanny theatre of fiction.

Dane, a famous writer, has difficulty finding an appropriate language to describe the Place. He often thinks of the Place with religious terminology:

> This recalled disposition of some great abode of an Order, some mild Monte Cassino, some Grande Chartreuse more accessible, was his main term of comparison; but he knew he had really never anywhere beheld anything at once *so calculated and so generous*.[74]

The Place, as it appears to Dane, has been designed and created by a mind of a higher order than the builders of religious spaces within the 'real' world of his everyday life. Dane later compares the Place to 'the bright country-house', a hotel and a club, only to

---

71 Salmon, *Henry James and the Culture of Publicity*, 176.
72 McGurl, *The Novel Art*, 69.
73 Ibid., 69–70.
74 H. James, *Novels and Tales*, 16: 236, emphasis added.

conclude that none of these comparisons is accurate enough. The 'only approach to a real analogy', Dane decides, is through himself and the other inhabitants of the Place, who are all 'made' by the conditions of the Place.[75] Dane and his companions seem to be wandering through the mind of their creator:

> What underlay and overhung it all, better yet, Dane mused, was some original inspiration, [...] some happy thought of an individual breast. [...] The author might remain in the obscure, for that was part of the perfection [...]. Yet the wise mind was everywhere – the whole thing was infallibly centred at the core in a consciousness.[76]

The consciousness here, which can be read as James's, is the negative space that lies at the centre of so many Jamesian texts; it is also like the Persian king, a creative 'void' that drives all life in the valley. The Place itself is figured in terms of gaps and absence: 'it was such an abyss of negatives, such an absence of positives and of everything', and for Dane, the escape from the 'real' world of the text in which he is figured provides the opportunity to just *be* there, to 'be without the complication of an identity'. 'Those things', Dane thinks 'were in the world'; thus, the Place is clearly not in the same 'world'.[77]

Another indication that the transcendental space of the Place is linked to the authorial consciousness is the higher level of communication that is possible in the Place; it is a kind of telepathy as Dane observes, 'established by the mere common knowledge of' the Place.[78] In fact, the inability to satisfactorily define or describe the Place seems to be a symptom of this heightened level of communication, between not only Dane and his companions, but between their minds and the physical reality of the Place. They have transcended spoken language; it is no longer necessary as an intermediary between subjective and objective realities. Sara Chapman writes that the Place is 'best understood as an extended metaphor of the writer in an ideal relation to the world around him'[79] and, indeed, Dane observes that in this 'scene of new consciousness': 'It was part of the whole impression that, by some extraordinary law, one's vision seemed less from the facts than the facts from one's vision; that the elements were determined at the moment by the moment's need or the moment's sympathy'.[80] This 'extraordinary law' is an expression of aesthetic will-to-power. In his argument with Wells, James claimed that 'it is art that *makes* life' and, here, in the Place, the reader is offered a literal example of the construction of 'reality' through the mediation of consciousness.

This Place, therefore, functions as a spatial representation of James's aesthetic impulse, 'a kind of fourth dimension' for his fictional character, Dane. Dane's exposure to the Place is figured as a mystical experience wherein he is able to recover his own creative impulse as a writer, of which he 'had a private practical sign [... ,] "the vision and the faculty divine"'.[81] Like Fleda, Dane's 'intense adventure of consciousness' grants him a

---

75   Ibid., 249.
76   Ibid., 250.
77   Ibid., 250.
78   Ibid., 237.
79   Chapman, *Henry James's Portrait of the Writer*, 108.
80   H. James, *Novels and Tales*, 16: 254.
81   Ibid., 248.

sense of personal autonomy: 'He had talked of independence and written of it, but what a cold flat word it had been! This was the wordless fact itself – the uncontested possession of the long sweet stupid day'.[82] Here, again, the issue of time arises; the day is only open to total possession from within the Place, where time has been abolished.

Henry James and 'Aesthetic Time'

Time is often reduced to 'nothing' in James, in that he tends to make it subordinate to space. As Georges Poulet observes, James 'invents a new kind of time, what one might call aesthetic time', where 'time is constituted by passage, not from one moment to another, but from one point of perspective to another'.[83] Similarly, Levin writes that often James 'depict[s] various aspects of the action [in his texts] in a kind of stopped time'.[84] James called this his 'dramatic principle', his 'law of successive aspects', and it corresponds to the 'fugitive nows' of Hinton's four-dimensional theory.[85] In James's perspective of his texts, the author is able to observe all possible moments of time within the represented world of his fiction simultaneously because he occupies the space outside and 'above' that of the text. The movement of the reader of the Jamesian text – both the external 'real-world' reader and the central consciousness – is across perspectives or aspects, engendering this 'aesthetic time' or, as Levin describes it, 'the temporal dynamic of [James's] own unfolding prose'.[86] Duration is thus an effect of movement across space.

In his polemical review of *The Wings of the Dove*, J. P. Mowbray attacks James's style in a manner that anticipates George Boon's. Mowbray struggled with James's 'aesthetic time', complaining of the novel's

> interminable and indeterminate dialogue, where indeed Mr. James seems to have reached the fourth dimension of space – dialogue in which the speakers not only tell us what they think and what others think, but what they might have thought and did n't.[87]

Mowbray used a spatial conception of the fourth dimension, one that places James in a position of omniscience where he is able to view all the possible outcomes of any given situation. Again, this is a space that lies outside, or above, the 'real', representational space of the novel. Mowbray's implication is that in his later style, James envisioned not only what *did* happen in a character's life within the narrative frame, but also the multitude of alternatives that *did not* happen within that frame, but perhaps could occur somewhere else. Mowbray, like Wells/Boon, complained that, as a result, James damaged the mimetic quality of his fiction:

---

82 Ibid., 251.
83 Poulet, *Studies*, 352.
84 Levin, *The Poetics of Transition*, 127.
85 H. James, *The Ivory Tower*, 276.
86 Ibid., 127.
87 Mowbray, 'The Apotheosis', 381.

He is so apprehensive when dealing with one shade of thought or emotion that there may be other subshades that he will miss and that he must clutch as he passes, that he frequently produces the effect of a painting niggled and teased out of all frankness by manipulation, and this, as we have already said, belongs as a method rather to chemistry than to art.[88]

What Mowbray observed here was James's attempt to achieve a 'view of *all* the dimensions' not only for himself, but his reader as well. Just as, according to Elizabeth Helsinger, 'reading Ruskin can be learning to see with Ruskin', to read James 'successfully' is to learn to read with James: a 'corrective lens' is necessary to bring his work into focus.

Playing on the multiple meanings of 'anamorphosis' in her examination of the influence of Hans Holbein the Younger's painting, *The Ambassadors* (1533), on James's novel of the same name, Richardson observes:

> Strether's preoccupation with the idea of how he could have lived [...] – effectively a *memento mori* – distorts the values with which he arrived in the Old World and at the same time marks a 'progression or change in form from one type to another [...]'. At bottom, beneath these variants, is the fact that Strether is 'formed anew' in and by his experiences as ambassador; there is a change in the 'organ' of his consciousness. He is anamorphic. Following the turns of his consciousness as it is presented by James, our consciousness is shaped into the 'proper restoring device'.[89]

Like a trained observer of the Holbein portrait, the Jamesian reader can register both the aesthetic and the ethic of the text, the Lambinet impression and the 'moral dimension'.

To the uninitiated, the effect of viewing a three-dimensional representation of the fourth dimension results in a vision so 'teased out of all frankness by manipulation' that it does not appear to be coherent or representative of reality. However, for one who possesses 'a view of *all* the dimensions', such a text is representative of a higher, 'more real', reality. This is an aesthetic that transcends the binary opposition of beauty and morality, of – to borrow Ruskin's terms – 'aesthesis' and 'theoria'. The integration of these values results in a reconfiguration of identity in the perceiving subject. No longer driven by the simple pursuit of pleasure or avoidance of pain, the Jamesian central character becomes, like the student of 'The Persian King', a creative agent in his or her own right. These subjects possess a sense of untimeliness, one which is usually gained – like the narrator of *An Unfinished Communication* – through revisiting and revising the past.

'The Jolly Corner' (1908)

Mowbray's complaint was that, in his later style, James attempted to depict not only what happened in any given situation, but also what did *not* happen, and we find that

---

88 Ibid., 380.
89 Richardson, *A Natural History*, 156–57.

this idea is perhaps most explicitly worked out in 'The Jolly Corner'. The protagonist, Spencer Brydon, is concerned with the seemingly irrevocable nature of the past. It is only through his uncanny experiences resulting from a return to his childhood home that Brydon is able to resolve his sense of 'something dreamed and missed, something reduced, relinquished, resigned', as Fleda would say. This space of his past, the 'house on the jolly corner', becomes a place of aesthetic time, a kind of authorial, four-dimensional consciousness where Brydon encounters his past as well as an alternative present, imagined as his 'other self'.

Brydon, like the narrator of *An Unfinished Communication*, is concerned with the way certain decisions in his past have narrowed and circumscribed his present situation. Unlike Hinton's narrator, he is not in a state of desperation at the beginning of the story; however, the narrative develops in such a way that he is finally brought to a similar crisis point. Returning to his childhood home in New York after 33 years in Europe, Brydon becomes obsessed with the idea of the man he could have been, or would have been, had he remained at home in the United States. Like Fleda, who can sense 'a kind of fourth dimension' in the maiden aunt's house, Brydon tells his companion, Alice Staverton, in reference to the 'ghostly' empty house of his childhood years: '"For me it *is* lived in. For me it *is* furnished." At which it was easy for her to sigh, "Ah yes – !" all vaguely and discreetly; since his parents and his favourite sister, to say nothing of other kin, in numbers, had run their course and met their end there'.[90] While Alice appears to be thinking of Brydon's deceased family, it is clear that – in Brydon's imagination – the current occupant of the house is in fact another version of himself, his 'Americanized' self. The narrator, aligned with Brydon's point of view, distinguishes the presence of this 'other' self from Alice's possible allusion to his deceased family, observing that *they* 'represented, within the walls, ineffaceable life'.[91] The implication here is that – at this point, early in the story – while Brydon respects the autonomy of the members of his family, he thinks of his other self as effaceable.

What begins as curiosity about 'the road not taken' becomes an obsession for Brydon as he spends more time in New York, making nightly visits to his empty childhood home. The reality of the other self is created in Brydon's imagination through a process of analogy; thus, we see the origin of the other self, as Hocks observes, in

> the incantatory repetition in the [...] lines – 'If he had but stayed at home' – as well as James's expression that Brydon's condition begins first as an 'analogy' which eventually he is to 'improve on' by a 'still intenser form,' makes clear at which end the process for reaching the quasi-supernatural must start from.[92]

It is this obsession with an unactualized past that leads Brydon to construct the vision of his alter ego near the end of the story, a vision so real that both Brydon and Alice experience it, and Brydon nearly dies from the shock of it.

---

90 H. James, *Novels and Tales*, 17: 447, original emphasis.
91 Ibid., 477.
92 Hocks, *Henry James and Pragmatistic Thought*, 202, emphasis removed.

It is significant that both Alice and Brydon 'see' the alter ego in separate instances, which, Hocks argues, demonstrates 'James's development and grasp of the ghostly realm, [that] he now seems less doubtful of its possible "independent" existence in his decision to allow both characters to participate in it'.[93] While Dane has shadowy companions in the Place, it remains ambiguous as to whether these 'Brothers' are just other aspects of himself, his servant or the young writer in his study. Dane's 'adventure of consciousness' is a solitary one, but both Brydon and Alice interact with this 'ghost' of Brydon's other self. Thus, in 'The Jolly Corner' the 'ghostly realm', or space, or state of consciousness, has a more stable existence, one that may originate within a creative consciousness but is also able to transgress the boundaries of that single consciousness. The alter ego, invented via analogy within Brydon's mind, has been externalized and reified into an entity that is visible to Alice as well.

Earlier, I explored the possibilities of reading the 'quasi-supernatural' spaces of *The Spoils* and 'The Great Good Place' as places where the Jamesian centre of consciousness is able to sense the authorial presence of James. For a Jamesian character, access to the authorial viewpoint can be described using the dimensional analogy in which a two-dimensional being is 'lifted' up to the third dimension of space. This omniscient view is one available to the author, and – as we have seen in the case of 'The Great Good Place' – occasionally James brought one of his centres of consciousness into contact with his own authorial consciousness. Such an experience for a Jamesian character was often figured in terms of the supernatural, but perhaps a more accurate description would be hyperreal.

The perspective of James the author, 'a view of *all* the dimensions', would affect the Jamesian character as an intense, superhuman experience because, as Stephen Donadio observes:

> As James conceives the novel, the author's comprehensive awareness serves as the larger frame within which the account of the more or less unsuccessful efforts of individual characters to unravel the tangle of their lives is finally placed. As a consequence, there is a continuous implicit comparison between the limited 'points of view' of particular persons and what may be characterized as the impersonal, all-inclusive point of view, which does not simply exemplify one of the degrees of lucidity represented as humanly possible by the story – for it lies always just beyond the reach of the capacities of even the most intense perceiver – but which remains at all times an unattainable ideal of total clarity and refinement of perception.[94]

However, this 'all-inclusive point of view' is not strictly impersonal for all of James's characters, such as Fleda Vetch and George Dane. These intense moments of awareness that they experience, of the 'life' beyond their narrative frame, allow the central consciousness to, as the narrator of *An Unfinished Communication* does, return from 'this wide view, and plung[e]' back into the business of life. Dane returns to his study, to his life as a well-known and respected author, just as the final words of *The Spoils of Poynton* are Fleda's

---

93 Ibid., 205.
94 Donadio, *Nietzsche, Henry James and the Artistic Will*, 150–51.

declaration, after witnessing the destruction of Poynton, that, 'I'll go back'.[95] Though these encounters with the Jamesian consciousness allow the central characters to become artists in their own right (one of those agents who '*makes* life', as James would say), the experience is apparently too intense for some readers. Mowbray complained that, after reading *The Wings of the Dove* within a 24-hour sitting, upon returning to the 'real' world outside of the text and James's consciousness, 'We shall be pardoned if we feel like the resuscitated man who, after being rolled on a barrel, is expected to reveal something of the mysterious midway between this and another state of existence'.[96] Like Fleda and Dane, Mowbray struggled to find an adequate language to describe the experience of immersion in the Jamesian consciousness which, he facetiously implied, can feel similar to a life-threatening experience.[97]

Daniel Mark Fogel highlights a notebook entry from 1895, where James first speculates on the possible 'germ' of a story that bears striking resemblance to 'The Jolly Corner'.[98] In this notebook entry, from 5 February 1895, James described an idea for a story upon the theme of 'too late', along the lines of Strether's character in *The Ambassadors*. However, James realized his idea that 'this Dead Self of the poor man's lives for him still in some indirect way' would make for a different narrative to Strether's: 'I've only to write a few words, however, to see that the 2 ideas have nothing to do with each other. They are different stories'.[99] In his original conception of what, Fogel argues, would become Brydon's encounter with his 'Dead Self', James foresaw that 'he himself, the man, must, *in* the tale, also materially die – die in the flesh as he had died long ago in the spirit, the *right* one. Then it is that his lost treasure revives most – no longer *contrarié* by his material existence'.[100] Brydon does not actually die in 'The Jolly Corner', however. After encountering his other self, he faints and is revived later by Alice, who momentarily believes him to be dead. Brydon also describes his swoon as a sort of death: '"It must have been that I *was* [dead]." He made it out as she held him. "Yes – I can only have died"'.[101] The intrusion of the other self into the space of the house – which functions as a kind of frame for Brydon's 'adventure of consciousness' – is too intense an experience even for a Jamesian centre of consciousness. The encounter between Brydon and his other self is figured an explosive disruption of his understanding of space and time.

Brydon must first overcome his denial of his other life, the roots of which already lie within his current life, his 'Europeanized self'. At the end of the story, it is through Alice that Brydon is able to acknowledge the unity between these two selves: although he explicitly denies their shared identity, he also implicitly accepts that Alice's occult vision

---

95 H. James, *The Spoils*, 208.
96 Mowbray, 'The Apotheosis', 379.
97 In a similar manner, Hiram Barton, a reader of *Scientific American*, wrote in to warn that Hinton's cube exercises (which he claims he 'nearly got hooked on [...] in the nineteenth-twenties') 'are completely mind-destroying' (Gardner, *Mathematical Carnival*, 52–53).
98 Fogel, 'A New Reading', 191.
99 *The Notebooks of Henry James*, ed. Matthiessen and Murdock, quoted in Fogel, 'A New Reading', 191.
100 Fogel, 'A New Reading', 192, original emphasis.
101 H. James, *Novels and Tales*, 17: 480, original emphasis.

of the other self is in fact *him*. In explaining her vision of the other self, the morning after Brydon's encounter with him, Alice conflates the two:

> 'Well, in the cold dim dawn of this morning I too saw you.'
> 'Saw *me* – ?'
> 'Saw *him* –,' said Alice Staverton. 'It must have been at the same moment.'
> 'Yes [...]. He came back to me. Then I knew it for a sign. He had come to you.'
> [...] '*He* didn't come to me.'
> 'You came to yourself,' she beautifully smiled.[102]

Unlike Brydon, Alice recognizes the latent qualities in the alter ego that link him to the version of Brydon that she knows as her friend: '*I* didn't disown him', she remarks.[103]

I am not inclined to read Brydon's acquiescence to Alice's embrace at the end of this story as either the 'happy' heterosexual ending lacking from 'The Beast in the Jungle'[104] or his nightmarish submission to compulsive heterosexuality,[105] but rather as his and Alice's 'shared recognition' of their otherness.[106] Hugh Stevens suggests that

> Brydon's relationship to his spectral double illustrates the ambivalence the late nineteenth-century homosexual subject feels towards a fantasy of masculinity as violent, sexually abusive, and powerful, as this fantasy represents, all at once, an erotic object (the allure of 'rough trade'), a possible figure of identification, and also, in its possible homophobia, a threat.[107]

While Brydon is terrified and repulsed by his alter ego, it is Alice who accepts, and even 'likes' him. As Stevens observes, she admits that she dreams of Brydon, but only of his doubled, 'Americanized' self. Alice sees and accepts both of his possible identities, assisting Brydon in re-assimilating both at the end of the tale.

In returning to his childhood home and confronting his alter ego, Brydon completes a cycle of return with variation. His return to consciousness in Alice's arms at the end of the story is a sort of rebirth, as Fogel observes, and Alice's final claim that 'he is n't – no, he is n't – *you*!' is not a straightforward denial of Brydon's other self.[108] 'Alice's underlying message entails a shift in tense: the apparition was you, both as you might have been and as you were, but he is not you now'.[109] Thus, the growth of Brydon's consciousness in this story follows the pattern of an expanding spiral similar to that of the narrator in *An Unfinished Communication* whose story ends with his acceptance of his other self at the dawn of a similar kind of 'rebirth'.

---

102 Ibid., 485, original emphasis. See also Fogel, 'A New Reading'.
103 H. James, *Novels and Tales*, 17: 485, original emphasis.
104 See Krook, *The Ordeal*, 334.
105 See Savoy, 'The Queer Subject'.
106 Stevens, 'Homoeroticism', 137.
107 Ibid., 136.
108 H. James, *Novels and Tales*, 17: 485, original emphasis.
109 Fogel, 'A New Reading', 199.

In his examination of the James brothers and time in *Henry James and the Abuse of the Past*, Peter Rawlings observes that 'for Henry James, objects, and especially houses, are often attractive because they spatialize and inscribe time in ways that disrupt linear, chronological models'.[110] The house of Brydon's childhood, like the house that Ralph Pendrel inherits in *The Sense of the Past*, is a site where time seems to become more flexible and open to overwriting. However, while the untimely house in *The Sense of the Past* provides a site from which Pendrel is able to visit the eighteenth century, the house in 'The Jolly Corner' has a slightly different role to play. What happens in Brydon's house, and in the modern, twentieth-century New York to which he returns, is, as Rawlings observes, 'less a travelling back, or a simple reversal of a sequence, and more the giddying experience of simultaneity as the unreality of time'.[111] The other self that Brydon encounters inhabits a stream of time that is separate from his own; the catalyst for this divergence, it is implied in the story, was Brydon's departure for Europe 33 years earlier. Like Hinton's narrator, who is able to experience the simultaneous events of his past life as well as those of his own other parallel lives as 'actual presences', Brydon encounters the presence of his simultaneously developing alter ego. As a companion in the Place tells Dane, 'I don't speak of the putting off of one's self; I speak only – if one has a self worth sixpence – of the getting it back'.[112] Brydon does not put off his other self, but his confrontation and acceptance of it – via Alice – allow him to acknowledge, and even reclaim these seemingly latent aspects of himself. Both Dane and Brydon awake from their experiences with their consciousnesses enriched: they have become 'more' of who they were already.

In such experiences time *is* 'unreal'; it is rendered as space, where multiple temporal realities may exist alongside one another. Hinton's concept of the fourth dimension also implies that time is actually an illusion; it is the way that humans with three-dimensional sensibility perceive the fourth dimension of space, and each three-dimensional slice, or 'fugitive now', is actually just one aspect of a four-dimensional, 'higher reality'. Brydon's experiences within the house on the climactic night of his encounter with his other self are also compartmentalized: 'the house, as the case stood, admirably lent itself; he might wonder at the taste, [...] which could rejoice so in the multiplication of doors – '.[113] Brydon becomes obsessed with opening and closing these doors, pausing on thresholds and searching each open or closed doorway for clues to his alter ego's movements through the house.

In his own upward progression through the house, Brydon conscientiously leaves open each door through which he passes. Turning to descend back down the house, Brydon finds a door closed, and 'with it rose, as not before, the question of courage – for what he knew the blank face of the door to say to him was "Show us how much you have!" It stared, it glared back at him with that challenge; it put to him the two alternatives: should he just push it open or not?'.[114] Brydon decides that the closure of a door he remembers

---

110 Rawlings, *Henry James and the Abuse of the Past*, 140. See also Waters, 'Still and Still Moving'.
111 Rawlings, *Henry James and the Abuse of the Past*, 147.
112 H. James, *Novels and Tales*, 16: 240.
113 Ibid., 17: 466.
114 Ibid., 467.

to have left open must signify the presence of his other self in the house, and that the other self is waiting behind the closed door. Deciding in favour of discretion – out of fear for his own sanity – Brydon leaves the door unopened and hastily begins his descent of the four levels of the house.

Edel highlights the obvious autobiographical aspects of 'The Jolly Corner', claiming that in this story, James is 'laying the ghost' of his own past.[115] Without wishing to reduce this text to autobiographical self-therapy, I agree with Edel that

> the story is more than a revisiting of a personal past; it becomes a journey into the self, almost as if the house on 'the jolly corner' were a mind, a brain, and Spencer Brydon were walking through its passages finding certain doors of resistance closed to truths hidden from himself.[116]

Of one door in particular, the closed one that signals to Brydon the presence of his other self, 'he knew – yes, as he had never known anything – that, *should* he see the door open, it would all too abjectly be the end of him'.[117] Should the door open to reveal his alter ego, Brydon fears what William James would deem a moral 'failure' – his refusal to sustain a representation. In the horror of being confronted with the undeniable presence of his alter ego, Brydon worries he will throw himself out the fourth-floor window: 'He saw himself uncontrollably insanely fatally take his way to the street'.[118] Making his way from this door and down a flight of stairs, Brydon begins to see every door he encounters as a threat. It seems that the multiplication of spaces also implies the potential for the multiplication of parallel selves: 'The house, withal seemed immense, the scale of space again inordinate; the open rooms, to no one of which his eyes deflected, gloomed in their shuttered state like mouths of caverns'.[119] The space of the house has become like the consciousness of James, the author, where not only the events that *did* actually happen for a character are present, but also events that did *not* happen. Like Strether, who steps into the Lambinet painting, Brydon has entered a work of art – James's 'house of fiction' – which 'confuses the categories of internal and external by objectifying some deeply personal element of consciousness'.[120]

Brydon's decision at specifically this point in the narrative to have the house torn down is significant. He imagines 'the clear delight with which he was finally to sacrifice it. They might come in now, the builders, the destroyers'.[121] Such an act of ressentiment would demolish the very foundation of Brydon's existence; it would be not only the ultimate act of revenge upon himself – by denying his alter ego – but it would also be a symbolic attack upon his creator, James. It is important, therefore, that upon reaching the ground floor, Brydon finds that another door – one he is certain he previously closed – 'had been thrown far back'.[122] Feeling his 'ease increased with the sight of the old

---

115 Edel, *Henry James, The Master*, 316.
116 Ibid., 314.
117 H. James, *Novels and Tales*, 17: 417, original emphasis.
118 Ibid., 472.
119 Ibid., 472–73.
120 Hutchison, *Seeing and Believing*, 172.
121 H. James, *The Novels and Tales*, 17: 473.
122 Ibid., 474.

black-and-white slabs' of the tiled ground floor (which he later refers to as 'these marble squares of his youth') Brydon is finally confronted with his alter ego, who is waiting for him at the foundation of the house.[123]

The alter ego is at first hiding his face in his hands, presumably in anguish or shame, but he soon drops his hands and advances aggressively toward Brydon.[124] Lustig offers a reading of this encounter that is illuminating: 'Brydon's other self stands at an open door looking inwards from the outer threshold of the fiction and the house in order to trap the expanding or escaping subjectivity'.[125] This forced encounter with his other self, this trapping of 'the expanding or escaping subjectivity' is protective as well as restrictive: Brydon's unchecked rush out the front door of his childhood home could be as dangerous as the panicked jump from the window he had envisaged earlier, a fatal *passage a l'acte* as evacuation from the representational space of his own text.

The house in 'The Jolly Corner' functions as a sort of uncanny borderland space between the mimetic world of the text and James's consciousness as author, what D. H. Lawrence would identify as 'the much-debated fourth dimension'.[126] Like Dane, Brydon senses a presence somehow behind the scenes, closing doors that he left open and opening ones he left closed. It is significant that Brydon's climactic encounter with this other self is described as a penetration of boundaries: 'No portrait by a great modern master could have presented him with more intensity, *thrust him out of his frame* with more art'. The alter ego, Brydon's inverted self in all his 'queer actuality', has stepped outside of the frame and into the house of fiction he shares with Brydon.[127] Faced with 'a life larger than his own, a rage of personality before which his own collapsed', Brydon literally collapses. His individual personality is subsumed within the explosion of the encounter: 'His head went round; he was going; he had gone'.[128]

Brydon's awakening is figured as a convergence of the selves that have been dissipated throughout the very air of the house, like the settling of dust after an explosion:

> In this rich return of consciousness – the most wonderful hour, little by little, that he had ever known, leaving him, as it did, so gratefully, so abysmally passive, and yet as with a treasure of intelligence waiting all round him for quiet appropriation; dissolved, he might call it, in the air of the place and producing the golden glow of a late autumn afternoon.[129]

The collision of the parallel lives of Brydon and his alter ego appears to have resulted in the temporary dissolution of both. The Brydon that survives this encounter is a reconfiguration of the personalities of both selves, as implied by his acceptance of Alice's conflation of the two identities cited earlier. Immersed in these diffused presences, Brydon feels that he can 'quietly appropriate' these identities and intelligences at will; he is engaging

---

123 Ibid., 473 and 478.
124 Ibid., 477.
125 Lustig, *Henry James and the Ghostly*, 224.
126 Lawrence, *Study of Thomas Hardy*, 171.
127 H. James, *The Novels and Tales*, 17: 475, emphasis added.
128 Ibid., 477.
129 Ibid., 478.

in an act of self-creation. Although, as opposed to Brydon's dangerous encounter in the house, Dane's excursion to the Place is a wholly comforting experience, both Dane and Brydon feel that they are in deeper possession of themselves at the end of their narratives. Looking around at the 'golden glow' produced by the dissolution of his conflicting selves, Brydon feels that 'what he had come back *to* seemed really the great thing, and as if his prodigious journey had been all for the sake of it'.[130] Brydon has completed another arc of the spiral; he has come back around to himself, with 'a treasure of intelligence', or expanded consciousness.

'A Kind of Fourth Dimension' in Henry James

In my examinations of Wells and James, I have focused on their similar concerns with framing, the spiralling consciousness and reader response, while approaching their texts through the lens of Hinton's hyperspace philosophy. While Wells seemed to be concerned with raising reader awareness by overtly interrupting the process of reading via the splintering frame technique, James dramatized the process in which, as Levin argues, 'by virtue of a heightened attentiveness and responsiveness, the artist or novelist sees beyond isolated or apparently trivial details into a more comprehensive reality'.[131] Wells fantasized about the development of a race of Overminds, or superhuman intelligences, whose consciousness would relate to human intelligence much in the same way that higher-dimensional beings relate to lower-dimensional beings.

James had his own interest in highly developed intelligences; his centres of consciousness often brush against the consciousness of James, himself; these encounters are experienced by the characters as secular, yet spiritual and hyperreal. The 'demonic' intelligence of these centres of consciousness allows them to traverse the borderland between James's mind and the represented space within the text. In their encounters with the Jamesian presence, James's centres of consciousness develop a greater sense of autonomy, as if they have learned from their author the art of self-construction. Henry McDonald describes this tendency in the Jamesian centres of consciousness: 'The source of the "active" morality embodied in James's tragic heroines and heroes is the self-submitting, self-creating artistic action of James himself'.[132] The Jamesian protagonists I have examined here mimic the creative activity of James as author; these are characters who 'overcome' themselves to become more like James, with his heightened awareness and creative agency. This pattern of 'self-submitting' and 'self-creating' also mimics the pattern of the spiral, of Wells's return with variation.

If, as James wrote to Wells, 'it is art that *makes* life, makes interest, makes importance for our consideration', then each expansion of the frame of the work of art, each completed arc of the spiral is an act of self-overcoming for the Jamesian centre of consciousness.[133] McDonald similarly observes that 'James felt that we live in only to the degree

---

130 Ibid., 479, original emphasis.
131 Levin, *The Poetics of Transition*, 132.
132 McDonald, 'Henry James as Nietzschean', 405.
133 Edel and Ray, eds, *Henry James and H. G. Wells*, 267, original emphasis.

we "exceed" ourselves, only to the degree we invest our hearts and minds in the hearts and minds of others and undergo a process of revolutionary self-transformation'.[134] Here again we see a kind of transcendental materialism. The living, embodied self is not abandoned, but it is transformed through achieving a higher dimension of consciousness – one that is distinctly literary.

---

134 McDonald, 'Henry James as Nietzschean', 405.

AFTERWORD

In 1905, Einstein was supporting himself by working as a clerk at the Swiss patent office while completing his doctoral thesis, 'A New Determination of Molecular Dimensions', and preparing his paper on his special theory of relativity, 'On the Electrodynamics of Moving Bodies'. Across the Atlantic, Charles Howard Hinton was working at his day job in the US Patent Office and revising his cube exercises into 'A Language of Space' for publication with his second edition of *The Fourth Dimension*. Hinton died in 1907 and, by the middle of the twentieth century, he had become a footnote in the history of science and science fiction.

Studies of relativity theory and its impact on the arts tend to focus on Henri Bergson's philosophical writings as an intermediary between the two cultures. Hinton's work runs in parallel with Bergson's in many ways, although there are important key differences.[1] The translation of Bergson into English in the 1910s as well as the popularization of Einstein's work from 1919 onward no doubt contributed to the shift of attention away from Hinton's hyperspace philosophy. The fact that Hinton did not live long enough to incorporate Einstein's ideas into his own work, while Bergson did, was also a contributing factor.

William James's discovery of the writings of Henri Bergson and disavowal of the 'logic' of higher dimensionality in *A Pluralistic Universe* (1909) is indicative of this shift. Although in this text James frequently employed spatial metaphors such as Fechner's 'mother-sea' and described his belief in possibility of 'higher consciousnesses' in language that is very similar to Hinton's, he also stated:

> I prefer to call reality if not irrational then at least non-rational in its constitution, – and by reality here I mean reality where things *happen*, all temporal reality without exception. I find myself no good warrant for even suspecting the existence of any reality of a higher denomination than that distributed and strung-along and flowing sort of reality which we finite beings swim in. That is the sort of reality given to us, and that is the sort with which logic is so incommensurable. [...] I should not now be emancipated, not now subordinate logic with so very light a heart, [...] if I had not been influenced by a comparatively young and very original french [sic] writer, Professor Henri Bergson.[2]

---

1 Bergson's three major texts, *Time and Free Will* (1889), *Matter and Memory* (1896) and *Creative Evolution* (1907), appeared nearly in tandem with Hinton's major publications.
2 W. James, *Writings, 1902–1910*, 726, original emphasis.

The 'useful fiction' of higher space was no longer useful to James; Bergson's *la durée pure* provided a more immediate way of treating 'reality'.

There is no evidence that Hinton encountered Bergson's work, or that he was able to read French.[3] Had he encountered Bergson, it seems likely that Hinton would have found his concept of *la durée pure* deeply opposed to his own hyperspace philosophy. While Bergson shared Hinton's concern about the limitations of language and epistemology, Bergson saw the spatialization of time as the fundamental flaw in Western thought:

> When we make time a homogeneous medium in which conscious states unfold themselves, we take it to be given all at once, which amounts to saying that we abstract it from duration. This simple consideration ought to warn us that we are thus unwittingly falling back upon space, and really giving up time.[4]

Bergson argued that temporal movement, or duration, is heterogeneous, and thus cannot be broken into discrete units. Motion happens in time, and 'there are two elements to be distinguished in motion, the space traversed and the act by which we traverse it, the successive positions and the synthesis of these positions. The first of these elements is a homogeneous quantity: the second has no reality except in a consciousness'.[5] This is distinct from Hinton's 'permanent things and a moving consciousness' in an important way. In Bergson's philosophy, there are no 'permanent things'. The 'homogeneous quantity' of space is in fact imposed upon human intuition by the intellect. The only reality, for Bergson, is the 'strung-along and flowing sort of reality' that the logical mind chops into discrete units. Thus, while Hinton argued that duration is the way in which the mind encounters the fourth dimension of space, Bergson believed that applying spatialized logic to the experience of temporality was in fact artificial and limiting.

The past, according to Bergson, is not 'visitable': 'as if this localizing of a *progress* in space did not amount to asserting that, even outside consciousness, the past co-exists along with the present!'.[6] As we have seen, for many contemporary readers, the co-existence of the past *is* exactly what is implied by Hinton's hyperspace philosophy. The hyperreal fourth dimension oscillates between the surface of the simulacrum, and the *sui generis* of a Platonic transcendent 'reality'. It is the useful fiction that allows for the methodologies of 'ambulatory relations' and the 'Arabic method of description', both of which imply a spatialized way of thinking. The mind cannot 'move through' experiences, constantly reassessing and revising, unless these experiences are imagined to inhabit a 'space', just as a reader cannot fill in the 'gaps of indeterminacy' of a text unless the text inhabits the physical realm.

I have described Hinton's hyperspace philosophy as a kind of transcendental materialism, and throughout my discussion of the dimensional analogy and framing the

---

3 Bergson's work began appearing in English translation only after Hinton's death. *Time and Free Will* was first published in English in 1910. *Matter and Memory*, *Laughter* and *Creative Evolution* appeared in 1911.
4 Bergson, *Time and Free Will*, 98.
5 Ibid., 112.
6 Ibid., 112, original emphasis.

emphasis has been on *textual* nature of Hinton's project. At a fundamental level, Hinton's hyperspace philosophy is a celebration of the technology of representation. For Hinton, it is by pushing 'through' matter (or space, or the text) that one is able to access a higher dimension of consciousness. Like Mallarmé, Hinton drew our attention to the materiality of the text itself:

> I have often thought, travelling by railway, when between the dark underground stations the lads and errand boys bend over the scraps of badly printed paper, reading fearful tales – I have often thought how much better it would be if they were doing that which I may call 'communing with space'. 'Twould be of infinite delight, romance and interest; far more than are those creased tawdry papers, with no form in themselves or in their contents.
>
> And yet, looking at the same printed papers, being curious, and looking deeper and deeper into them with a microscope, I have seen that in splodgy ink stroke and dull fibrous texture, each part was definite, exact, absolutely so far and no further, punctiliously correct; and deeper and deeper lying a wealth of form, a rich variety and amplitude of shapes, that in a moment leapt higher than my wildest dreams could conceive. [...] For there, in these crabbed marks and crumpled paper, there, if you but look, is space herself, in all her indefinite determinations of form.[7]

To the refined aesthetic consciousness (one with an almost microscopic vision of minute details), a rich and beautiful world abounds between the lines of even the most crudely material of texts. For Hinton, the text functions as a threshold to the fourth dimension; it is through 'reading' that one enters it.

Unaware of Hinton's bigamy conviction and his resultant exile from England, Borges wrote in the preface to his translation of Hinton's romances:

> Hinton's *A New Era of Thought* (1888) includes a note from the editors which says: 'The manuscript that is the basis for this book was sent to us by its author, shortly before his departure from England for an unknown and remote destination.' [...] This suggests a probable suicide or, more likely, that our fugitive friend had escaped to the fourth dimension which he had glimpsed, as he himself told us, thanks to a steadfast discipline. [...] Why not suppose Hinton's book to be perhaps an artifice to evade an unfortunate fate? Why not suppose the same of all creators?[8]

Borges's observation that Hinton's escape to hyperspace was more likely than his 'probable' self-destruction nicely brings out the aesthetic potential of the fourth dimension before Einstein: it is the 'space' of literary self-creation.

---

7   Hinton, 'Many Dimensions', *Scientific Romances*, 33.
8   Borges, *The Total Library*, 510.

BIBLIOGRAPHY

Abbott, Edwin A. *Flatland: A Romance of Many Dimensions*. Oxford: Oxford University Press, 2006.
Ansell-Pearson, Keith. 'Who Is the *Übermensch*? Time, Truth, and Woman in Nietzsche'. *Journal of the History of Ideas* 52, no. 3 (1992): 309–31.
Arbib, Michael A. and Mary B. Hesse. *The Construction of Reality*. Cambridge: Cambridge University Press, 1986.
Armstrong, Karen. *The Battle for God: Fundamentalism in Judaism, Christianity and Islam*. London: Harper Collins, 2001.
Auden, W. H. *The Dyer's Hand and Other Essays*. London: Faber and Faber, 1962.
Bain, Alexander. *The Senses and the Intellect*. London: John W. Parker and Son, 1855.
Ballard, Marvin H. 'The Life and Thought of Charles Howard Hinton'. MA Thesis, Virginia Polytechnic Institute and State University (Virginia Tech), 1980.
Barthes, Roland. 'From Work to Text'. In *Modern Literary Theory: A Reader*, edited by P. Rice and P. Waugh, 191–97. London: Arnold Press, 1996.
Batchelor, John. *H. G. Wells*. Cambridge: Cambridge University Press, 1985.
Baudrillard, Jean. *Simulations*. Translated by P. Foss and P. Beitchman. Cambridge, MA: Massachusetts Institute of Technology Press, 1983.
———. *Selected Writings*. Edited by M. Poster. Cambridge: Polity Press, 1996.
Beaumont, Matthew. '"A Little Political World of My Own": The New Woman, the New Life and *New Amazonia*'. *Victorian Literature and Culture* 35, no. 1 (2007): 215–32.
Beer, Gillian. '"Authentic Tidings of Invisible Things": Vision and the Invisible in the Later Nineteenth Century'. In *Vision in Context: Historical and Contemporary Perspectives on Sight*, edited by T. Brennan and M. Jay, 85–98. New York and London: Routledge, 1996.
———. *Open Fields: Science in Cultural Encounter*. Oxford: Clarendon Press, 1996.
———. *Darwin's Plots: Evolutionary Narrative in Darwin, George Eliot and Nineteenth Century Fiction*. Cambridge: Cambridge University Press, 2000.
Bell, Ian F. A. 'The Real and the Ethereal: Modernist Energies in Eliot and Pound'. In *From Energy to Information: Representation in Science and Technology, Art, and Literature*, edited by B. Clarke and L.D. Henderson, 114–25. Stanford: Stanford University Press, 2002.
Bell, Ian F. A. and Meriel Lland. 'Silence and Solidity in Early Anglo–American Modernism: Nietzsche, the Fourth Dimension, and Ezra Pound, Part One'. *Symbiosis: A Journal of Anglo-American Literary Relations* 10, no. 1 (2006): 47–61.
———. 'Silence and Solidity in Early Anglo–American Modernism: Nietzsche, the Fourth Dimension, and Ezra Pound, Part Two'. *Symbiosis: A Journal of Anglo-American Literary Relations* 10, no. 2 (2006): 115–31.
Bell, Millicent. *Meaning in Henry James*. Cambridge, MA: Harvard University Press, 1991.
Bentley, Nancy. 'The Fourth Dimension: Kinlessness and African American Narrative'. *Critical Inquiry* 35, no. 2 (Winter 2009): 270–92.
Bergson, Henri. *Time and Free Will: An Essay on the Immediate Data of Consciousness*. Translated by F. L. Pogson. London: George Allen and Unwin, 1910.
———. *Introduction to Metaphysics*. Translated by T. E. Hulme. New York: G. P. Putnam's Sons, 1912.
Blacklock, Mark. *The Fairyland of Geometry: A Cultural History of Higher Space, 1853–1907* (blog). http://higherspace.wordpress.com/
———. 'Analogy and the Dimensional Menagerie'. *Interdisciplinary Science Reviews* 39, no. 2 (June 2014): 113–29.
Bohm, David and F. David Peat. *Science, Order, and Creativity*. London and New York: Routledge, 2000.

Bold, Paul. 'The Professor's Experiments'. *Idler: An Illustrated Monthly Magazine* (December 1910): 255–70.
Borges, Jorge Luis. *The Total Library: Non Fiction 1922–1986, Jorge Luis Borges*. Edited by E. Weinberger; translated by E. Allen, S. J. Levine and E. Weinberger. London: Penguin, 1999.
Bradbury, Nicola. *Henry James: The Later Novels*. Oxford: Clarendon Press, 1979.
Bragdon, Claude. *Four-Dimensional Vistas*. London: George Routledge and Sons, 1923.
Brake, Laurel. 'Degrees of Darkness: Ruskin, Pater and Modernism'. In *Ruskin and Modernism*, edited by G. Cianci and P. Nicholls, 48–66. Basingstoke and New York: Palgrave, 2001.
Brandon, Ruth. *The New Women and the Old Men: Love, Sex and the Woman Question*. London: Secker and Warburg, 1990.
Brent, Jason. *Charles Sanders Peirce: A Life*. Bloomington: Indiana University Press, 1998.
Brown, Bill. 'A Thing about Things: The Art of Decoration in the Work of Henry James'. *Henry James Review* 23, no. 3 (Fall 2002): 222–32.
Caldwell, Larry W. 'Time at the End of Its Tether: H. G. Wells and the Subversion of Master Narrative'. *Cahiers Victoriens et Edouardiens* 46 (October 1997): 127–43.
Carnap, Rudolph. 'Empiricism, Semantics, and Ontology'. In *Philosophy of Mathematics: Selected Readings*, edited by P. Benacerraf and H. Putnam, 241–57. Englewood, NJ: Prentice-Hall, 1964.
Carpenter, Edward. *From Adam's Peak to Elephanta: Sketches in Ceylon and India*. London: Swan Sonnenschein and Son, 1892.
Cartwright, Lisa. *Screening the Body: Tracing Medicine's Visual Culture*. Minneapolis and London: University of Minnesota Press, 1995.
Cassirer, Ernst. *Language and Myth*. Translated by S. K. Langer. New York: Dover, 1953.
Chapman, Sara S. *Henry James's Portrait of the Writer as Hero*. Basingstoke and London: Macmillan, 1990.
Clarke, Bruce. 'Allegories of Victorian Thermodynamics'. *Configurations* 4, no. 1 (1996): 67–90.
———. *Energy Forms: Allegory and Science in the Classical Era of Thermodynamics*. Ann Arbor: University of Michigan Press, 2001.
———. 'Dark Star Crashes: Classical Thermodynamics and the Allegory of Cosmic Catastrophe'. In *From Energy to Information: Representation in Science and Technology, Art, and Literature*, edited by B. Clarke and L. D. Henderson, 59–75. Stanford: Stanford University Press, 2002.
Culverhouse, Emily. 'Photography Up to Date – and Beyond It'. *International Annual of Anthony's Photographic Bulletin and American Process Yearbook, Vol. 9 for 1897* (1896): 75–78.
Daley, Kenneth. *The Rescue of Romanticism: Walter Pater and John Ruskin*. Athens: Ohio University Press, 2001.
Daugherty, Sarah B. 'James and the Ethics of Control: Aspiring Architects and Their Floating Creatures'. In *Enacting History in Henry James: Narrative, Power and Politics*, edited by G. Buelens, 61–74. Cambridge: Cambridge University Press, 1997.
Delbanco, Nicholas. *Group Portrait: Joseph Conrad, Stephen Crane, Ford Madox Ford, Henry James and H. G. Wells*. London: Faber and Faber, 1982.
Derrida, Jacques. 'Structure, Sign, and Play in the Discourse of the Human Sciences'. In *The Languages of Criticism and the Sciences of Man: The Structuralist Controversy*, edited by R. A. Macksey and E. Donato, 247–65. Baltimore and London: Johns Hopkins University Press, 1970.
———. *Of Grammatology*. Translated by G. Chakravorty Spivak. Baltimore: Johns Hopkins University Press, 1998.
Donadio, Stephen. *Nietzsche, Henry James and the Artistic Will*. New York: Oxford University Press, 1978.
Draznin, Yaffa C. *'My Other Self': The Letters of Olive Schreiner and Havelock Ellis, 1884–1920*. New York: Peter Lang, 1992.
Edel, Leon. *Henry James, he Master: 1901–1916*. Philadelphia and New York: J. B. Lippincott, 1972.
Edel, Leon and Gordon N. Ray, eds. *Henry James and H. G. Wells: A Record of Their Friendship, Their Debate on the Art of Fiction, and Their Quarrel*. London: Rupert Hart-Davis, 1958.

Einstein, Albert and Leopold Infeld. *The Evolution of Physics: The Growth of Ideas from Early Concepts to Quanta*. New York: Simon Schuster, 1938.
Ellis, Edith. *Three Modern Seers*. New York: Mitchell Kennerley, 1910.
Ellis, Havelock. *The Havelock Ellis Papers*, MSS ADD 750528, British Library.
———. 'Hinton's Later Thought'. *Mind* (July 1884): 384–405.
———. *Man and Woman: A Study of Human Secondary Sexual Characteristics*. London: Walter Scott, 1894.
———. *My Life*. London and Toronto: William Heinemann, 1940.
Ellis, Havelock, and John A. Symonds. *Sexual Inversion*. London: Wilson and MacMillan, 1897.
Emerson, Ralph Waldo. *Essays and Lectures*. New York: Library of America, 1983. 'Extraordinary Confession of Bigamy'. *Reynolds News*, 17 October 1886, n.p.
Fechner, Gustav Theodor. *Vier Paradoxa*. Leipzig: L. Voss, 1846.
———. *Elemente der Psychophysik*. Vol. 2. Leipzig: Breitkopf und Härtel, 1860.
———. *The Little Book of Life After Death*. Translated by M. C. Wadsworth. Boston: Little, Brown, 1907.
Fellows, Jay. *The Failing Distance: The Autobiographical Impulse in John Ruskin*. Baltimore: Johns Hopkins University Press, 1975.
Fogel, Daniel M. *Henry James and the Structure of the Romantic Imagination*. Baton Rouge and London: Louisiana State University Press, 1981.
———. 'A New Reading of Henry James's "The Jolly Corner" '. In *Critical Essays on Henry James: The Late Novels*, edited by J. W. Gargano, 190–203. Boston: G. K. Hall, 1987.
Ford, Ford Madox. *Mightier than the Sword: Memories and Criticisms*. London: George Allen and Unwin, 1938.
Forster, E. M. *Aspects of the Novel*. London: Edward Arnold & Co., 1937.
Frank, Joseph. *The Idea of Spatial Form*. New Brunswick and London: Rutgers University Press, 1991.
Freud, Sigmund. *The Basic Writings of Sigmund Freud*. Translated by A. A. Brill. New York: Modern Library, 1938.
Gardner, Martin. *Mathematical Carnival*. Harmondsworth: Penguin, 1982.
Gelley, Alexander. 'The Represented World: Toward a Phenomenological Theory of Description in the Novel'. *Journal of Aesthetics and Art Criticism* 37, no. 4 (Summer 1979): 415–22.
Gibbons, Tom. 'Cubism and "The Fourth Dimension" in the Context of Late-Nineteenth Century Occult Idealism'. *Journal of Warburg and Courtauld Institutes* 44 (1981): 130–47.
Glasser, Otto. *Dr. W. C. Röntgen*. Springfield, IL: Charles C. Thomas, 1958.
Gooding-Williams, Robert. *Nietzsche's Dionysian Modernism*. Stanford: Stanford University Press, 2001.
Green, T. H. *Works of Thomas Hill Green*. Edited by R. L. Nettleship. London: Longmans, Green, and Co., 1885–1888.
Greenberg, Clement. *Art and Culture: Critical Essays*. Boston: Beacon Press, 1989.
Grosskurth, Phyllis. *Havelock Ellis: A Biography*. London: Allen Lane, 1980.
Grosz, Elizabeth. *Space, Time, and Perversion: Essays on the Politics of Bodies*. London and New York: Routledge, 1995.
Hamilton-Gordon, E. A. 'The Fourth Dimension'. *Science Schools Journal* 5 (April 1887): 145–51.
Hardin, Michael. 'Ralph Ellison's *Invisible Man*: Invisibility, Race, and Homoeroticism from Frederick Douglass to E. Lynn Harris'. *Southern Literary Journal* 37, no. 1 (2004): 96–120.
Hardy, Sylvia. 'Wells the Poststructuralist'. *Wellsian* 16 (Summer 1993): 2–23.
Hayles, N. Katherine. *Chaos Bound: Orderly Disorder in Contemporary Literature and Science*. Ithaca and London: Cornell University Press, 1990.
———. 'Constrained Constructivism: Locating Scientific Inquiry in the Theater of Representation'. In *Realism and Representation: Essays on the Problem of Realism in Relation to Science, Literature, and Culture*, edited by G. Levine, 27–43. Madison: University of Wisconsin Press, 1993.
Helmholtz, Hermann von. 'The Axioms of Geometry'. *Academy* (February 1870): 128–31.
———. 'The Origin and Meaning of Geometrical Axioms'. *Mind* (July 1876): 301–21.

Helsinger, Elizabeth K. *Ruskin and the Art of the Beholder*. Cambridge and London Harvard University Press, 1982.
Henderson, Andrea. 'Math for Math's Sake: Non-Euclidean Geometry, Aestheticism, and *Flatland*'. *PMLA* 124, no. 2 (2009): 455–71.
Henderson, Linda Dalrymple. *The Fourth Dimension and Non-Euclidean Geometry in Modern Art*. Cambridge, MA, and London: Massachusetts Institute of Technology Press, 2013.
———. 'X Rays and the Quest for Invisible Reality in the Art of Kupka, Duchamp, and the Cubists'. *Art Journal* 47, no. 4 (1988): 323–40.
———. 'Four-Dimensional Space or Space-Time? The Emergence of the Cubism-Relativity Myth in New York in the 1940s'. In *The Visual Mind II*, edited by M. Emmer, 349–97. Cambridge, MA: Massachusetts Institute of Technology Press, 2005.
Herbermann, Charles G. et al., eds. *The Catholic Encyclopedia: An International Work of Reference on the Constitution, Doctrine, Discipline, and History of the Catholic Church*. London: Caxton Publishing, 1907–1914.
Herbert, Christopher. *Victorian Relativity: Radical Thought and Scientific Discovery*. Chicago and London: University of Chicago Press, 2001.
Hilton, Tim. *John Ruskin: The Later Years*. New Haven and London: Yale University Press, 2000.
Hinton, Charles Howard, *Correspondence with William Sonnenschein*. Records of George Allen & Unwin, Ltd, MSS 382, Book 28, Archive and Manuscript Division, University of Reading Special Collections.
———. *A New Era of Thought*. London: Swan Sonnenschein, 1888.
———. *Stella and An Unfinished Communication: Studies of the Unseen*. London: Swan Sonnenschein, 1895.
———. *The Fourth Dimension*. London: Allen and Unwin, 1906.
———. *A Language of Space*. London: Sonnenschein, 1906.
———. *An Episode of Flatland: Or, How a Plane Folk Discovered the Third Dimension*. London: Swan Sonnenschein, 1907.
———. *Scientific Romances, First Series and Second Series*. New York: Arno Press, 1976.
Hinton, James. 'On the Basis of Morals'. *Contemporary Review* (December 1875): 780–90.
———. *Chapters on the Art of Thinking and Other Essays*. Edited by C. H. Hinton. London: C. Kegan Paul & Co., 1879.
———. *The Law-Breaker and the Coming of the Law*. Edited by M. Hinton. London: Kegan Paul, Trench and Co., 1884.
———. *The Life and Letters of James Hinton*. Edited by E. Hopkins. London: Kegan Paul, Trench, Trübner and Co., 1906.
Hinton, Mary Boole. *Other Notes*. Washington, DC: Neale Publishing, 1901.
Hocks, Richard A. *Henry James and Pragmatistic Thought: A Study of the Relationship between the Philosophy of William James and the Literary Art of Henry James*. Chapel Hill: University of North Carolina Press, 1974.
Howells, William Dean. *My Literary Passions: Criticism and Fiction*. New York: Harper, 1895.
Hutchison, Hazel. *Seeing and Believing: Henry James and the Spiritual World*. New York and Basingstoke: Palgrave Macmillan, 2006.
Huxley, Thomas Henry. 'Scientific Education: Notes of an After-Dinner Speech'. *Macmillan's Magazine*, June 1869.
Iser, Wolfgang, 'Indeterminacy and the Reader's Response in Prose Fiction'. In *Aspects of Narrative: Selected Papers from the English Institute*, edited by J. Hillis Miller, 1–45. New York and London: Columbia University Press, 1971.
———. *The Implied Reader: Patterns of Communication in Prose Fiction from Bunyan to Beckett*. Baltimore and London: Johns Hopkins University Press, 1974.
———. *The Act of Reading: A Theory of Aesthetic Response*. London and Henley: Routledge and Kegan Paul, 1978.

Jackson, Holbrook. *The Eighteen Nineties: A Review of Art and Ideas at the Close of the Nineteenth Century.* London: Grant Richards, 1913.
James, Henry. *The Novels and Tales of Henry James.* New York: Charles Scribner's Sons, 1909.
———. *The Art of the Novel: Critical Prefaces.* New York and London: Charles Scribner's Sons, 1934.
———. *The Ivory Tower.* New York: Charles Scribner's Sons, 1945.
———. *The Spoils of Poynton.* London: Bodley Head, 1967.
———. *Henry James: Letters, 1895–1916.* Edited by L. Edel. Cambridge, MA: Harvard University Press, 1984.
———. *The Ambassadors.* Oxford: Oxford University Press, 2008.
James, William, *Letters from Various Correspondents*, bMS Am 1092, Houghton Library, Harvard University.
———. *The Principles of Psychology, Authorized, Unabridged Edition in Two Volumes.* New York: Dover, 1950.
———. *The Will to Believe and Other Essays in Popular Philosophy.* Edited by F. Burkhardt, F. Bowers and I. K. Skrupskelis. Cambridge, MA: Harvard University Press, 1978.
———. *Writings, 1902–1910.* New York: Library of America, 1987.
———. *Manuscript Lectures.* Cambridge, MA: Harvard University Press, 1988.
———. *Writings, 1878–1899.* New York: Library of America, 1992.
Jann, Rosemary. 'Abbott's "Flatland": Scientific Imagination and "Natural Christianity"'. *Victorian Studies* 28, no. 3 (Spring 1985): 473–90.
Jenkins, Alice, *Space and the 'March of Mind': Literature and the Physical Sciences in Britain 1815–1850.* Oxford: Oxford University Press, 2007.
Jordanova, Ludmilla. *Sexual Visions: Images of Gender in Science and Medicine Between the Eighteenth and Twentieth Centuries.* Hertfordshire: Harvester Wheatsheaf, 1989.
Jubin, Brenda. '"The Spatial Quale": A Corrective to James's Radical Empiricism'. *Journal of the History of Philosophy* 15, no. 2 (1977): 212–16.
Kant, Immanuel. *Prolegomena to Any Future Metaphysics That Will Be Able to Present Itself as a Science.* Translated by P. G. Lucas. Manchester: Manchester University Press, 1953.
Kern, Stephen. *The Culture of Time and Space: 1880–1918.* Cambridge and London: Harvard University Press, 2003.
Kevles, Bethann Holtzmann. *Naked to the Bone: Medical Imaging in the Twentieth Century.* New Brunswick: Rutgers University Press, 1997.
Koven, Seth. *Slumming: Sexual and Social Politics in Victorian London.* Princeton: Princeton University Press, 2004.
Krook, Dorothea. *The Ordeal of Consciousness in Henry James.* Cambridge: Cambridge University Press, 1962.
Land, Jan Pieter Nichols. 'Kant's Space and Modern Mathematics'. *Mind* (July 1876): 38–46.
Lapore, Jill. *The Secret History of Wonder Woman.* New York: Knopf, 2014.
Lawrence, D. H. *Study of Thomas Hardy and Other Essays.* Edited by B. Steele. Cambridge: Cambridge University Press, 1985.
Lears, T. Jackson. *No Place of Grace: Antimodernism and the Transformation of American Culture, 1880–1920.* Chicago and London: University of Chicago Press, 1994.
Lecercle, Jean-Jacques. *Philosophy of Nonsense: The Intuitions of Victorian Nonsense.* London and New York: Routledge, 1994.
Le Poidevin, Robin. *Travels in Four Dimensions: The Enigmas of Space and Time.* Oxford: Oxford University Press, 2003.
Levin, Jonathan. *The Poetics of Transition: Emerson, Pragmatism, and American Literary Modernism.* Durham, NC, and London: Duke University Press, 1999.
Levine, George. *Dying to Know: Scientific Epistemology and Narrative in Victorian England.* Chicago and London: University of Chicago Press, 2002.

Lewes, George Henry. 'Imaginary Geometry and the Truth of Axioms'. *Fortnightly Review* (August 1874): 192–200.
Lewis, Wyndham. *Time and Western Man*. New York: Harcourt, Brace, and Company, 1927.
Lockyer, William J. S. 'A Contribution to the New Photography'. *Nature* 53 (6 February 1896): 324.
Lustig, Timothy. *Henry James and the Ghostly*. Cambridge: Cambridge University Press, 1994.
MacKenzie, Norman Ian and Jeanne MacKenzie. *The First Fabians*. London: Weidenfeld and Nicholson, 1977.
Mann, Thomas. *The Magic Mountain*. Translated by H. T. Lowe-Porter. London: Vintage, 1999.
Manning, Henry Parker, ed. *The Fourth Dimension Simply Explained: A Collection of Essays Selected from Those Submitted in the Scientific American's Prize Competition*. New York: Munn and Co., 1910.
Marcus, Laura and Peter Nicholls, eds. *The Cambridge History of Twentieth-Century Literature*. Cambridge: Cambridge University Press, 2004.
Marshall, M. E. 'William James, Gustav Fechner, and the Question of Dogs and Cats in the Library'. *Journal of the History of Behavioral Sciences* 10, no. 3 (July 1974): 304–12.
Matthiessen, F. O., ed. *The James Family: Including Selections from the Writings of Henry James, Senior, William, Henry, and Alice James*. New York: Vintage Books, 1980.
Matthiessen, F. O. and Kenneth B. Murdock, eds. *The Notebooks of Henry James*. New York: Oxford University Press, 1947.
Maxwell, James Clerk. *The Scientific Letters and Papers of James Clerk Maxwell*, Vol. 2. Edited by P. M. Harman. Cambridge and New York: Cambridge University Press, 1995.
McDonald, Henry. 'Henry James as Nietzschean: The Dark Side of the Aesthetic'. *Partisan Review* 56, no. 3 (1989): 391–405.
McGurl, Mark. *The Novel Art: Elevations of American Fiction after Henry James*. Princeton and Oxford: Princeton University Press, 2001.
McLean, Steven. *The Early Fiction of H. G. Wells: Fantasies of Science*. Basingstoke and New York: Palgrave Macmillan, 2009.
McMurray, William. 'Reality in James's "The Great Good Place"'. *Studies in Short Fiction* 14, no. 1 (1977): 82–83.
Meisel, Perry. 'Psychoanalysis and Aestheticism'. *American Imago* 58, no. 4 (2001): 749–66.
Meyer, Steven. *Irresistible Dictation: Gertrude Stein and the Correlations of Writing and Science*. Stanford: Stanford University Press, 2003.
Michie, Helena. *The Flesh Made Word: Female Figures and Women's Bodies*. New York and Oxford: Oxford University Press, 1987.
Mitchell, L. C. '"To Suffer like Chopped Limbs": The Dispossessions of *The Spoils of Poynton*'. *Henry James Review* 26, no. 1 (2005): 20–38.
Morgan, John F. *The Invisible Man: A Self-Help Guide for Men with Eating Disorders, Compulsive Exercise and Bigorexia*. London and New York: Routledge, 2008.
Mould, Richard A. *A History of X-rays and Radium with a Chapter on Radiation Units: 1895–1937*. London: I. P. C. Business Press, 1980.
Mowbray, J. P. 'The Apotheosis of Henry James'. In *Henry James: The Contemporary Reviews*, edited by K. J. Hayes, 376–82. Cambridge and New York: Cambridge University Press, 1996.
Murphy, Patricia. *Time Is of the Essence: Temporality, Gender and the New Woman*. Albany: State University of New York Press, 2001.
Murray, Paul. *A Fantastic Journey: The Life and Literature of Lafcadio Hearn*. Folkstone: Japan Library, 1993.
Newman, Ernest. 'Oscar Wilde: A Literary Appreciation'. In *Oscar Wilde: The Critical Heritage*, edited by K. Beckson, 231–36. London: Routledge and Kegan Paul, 1970.
Nietzsche, Friedrich. *Also Sprach Zarathustra: Ein Buch für Alle und Keinen*. Leipzig: C. G. Naumann, 1904.
———. *Thus Spake Zarathustra*. London: J. M. Dent, 1950.

———. *Untimely Meditations*. Edited by D. Breazeale. Translated by R. J. Hollingdale. Cambridge: Cambridge University Press, 1997.
———. *The Birth of Tragedy and Other Writings*. Edited by R. Geuss and R. Speirs. Translated by R. Speirs. Cambridge: Cambridge University Press, 1999.
———. *Beyond Good and Evil: Prelude to a Philosophy of the Future*. Translated by R. J. Hollingdale. London: Penguin, 2003.
Nordau, Max Simon. *Degeneration*. Lincoln: University of Nebraska Press, 1993.
'Note'. *Nature* 53 (20 February 1896): 308.
'Notices'. *Nature* 53 (16 January 1896): 253.
Nottingham, Chris. *The Pursuit of Serenity: Havelock Ellis and the New Politics*. Amsterdam: University of Amsterdam Press, 1999.
Oakeshott, Joseph F. 'Practical versus Idealist Socialism'. *Seed-Time: The Organ of the New Fellowship* 4 (April 1890): 13–14.
O'Gorman, Francis. 'Ruskin and Particularity: *Fors Clavigera* and the 1870s'. *Philological Quarterly* 79, no. 1 (Winter 2000): 119–36.
Papin, Liliane. 'This Is Not a Universe: Metaphor, Language, and Representation'. *Proceedings of the Modern Language Association* 107, no. 5 (1992): 1253–65.
Payne, J. F. 'James Hinton'. *Mind* (April 1876): 247–52.
Peard, George. 'The Hintons: Father and Son'. *Contemporary Review* (May 1878): 259–71.
Perry, Ralph Barton. *The Thought and Character of William James, as Revealed in Unpublished Correspondence and Notes, Together with His Published Writings*. Boston and Toronto: Little, Brown and Company, 1935.
Plücker, Julius. 'On the New Geometry of Space'. *Proceedings of the Royal Society of London* 14 (1865): 53–58.
'Police'. *Times* (London), 16 October 1886.
Porter, Theodore M. *Karl Pearson: The Scientific Life in a Statistical Age*. Princeton: Princeton University Press, 2005.
Potts, Annie. *The Science/Fiction of Sex: Feminist Deconstruction and Vocabularies of Heterosex*. London and New York: Routledge, 2002.
Poulet, Georges. *Studies in Human Time*. Translated by E. Coleman. Baltimore: Johns Hopkins University Press, 1956.
Pound, Ezra. *The Cantos of Ezra Pound*. London: Faber and Faber, 1975.
Raknem, Ingvald. *H. G. Wells and His Critics*. Trondheim: Universitesforlaget, 1962.
Rampley, Matthew. *Nietzsche, Aesthetics and Modernity*. Cambridge: Cambridge University Press, 2000.
Rawlings, Peter. *Henry James and the Abuse of the Past*. Basingstoke: Palgrave Macmillan, 2005.
'Recent Science. Röntgen's Rays'. *Nineteenth Century* 39 (March 1896): 416–25.
Richards, Joan. *Mathematical Visions: The Pursuit of Geometry in Victorian England*. San Diego and London: Academic Press, 1988.
Richardson, Joan. *A Natural History of Pragmatism: The Fact of Feeling from Jonathan Edwards to Gertrude Stein*. Cambridge: Cambridge University Press, 2006.
Rucker, Rudy, ed. *Speculations on the Fourth Dimension: The Selected Writings of Charles H. Hinton*. New York: Dover, 1980.
Ruskin, John. *The Complete Works of John Ruskin*. Edited by E. T. Cook and A. Wedderburn. London: George Allen, 1903–1912.
———. *The Diaries of John Ruskin: 1874–1889*. Edited by J. Evans and J. H. Whitehouse. Oxford: Clarendon Press, 1959.
Rylance, Rick. *Victorian Psychology and British Culture 1850–1880*. Oxford: Oxford University Press, 2000.
Saiber, Arielle and Henry S. Turner. 'Mathematics and the Imagination: A Brief Introduction'. *Configurations* 17, no. 1 (2009): 1–18.

Salmon, Richard. *Henry James and the Culture of Publicity*. Cambridge: Cambridge University Press, 1997.
Savoy, Eric. 'The Queer Subject of "The Jolly Corner."' *Henry James Review* 20, no. 1 (1999): 1–21.
Scheick, William J. *The Splintering Frame: The Later Fiction of H. G. Wells*. Victoria, BC: University of Victoria Press, 1984.
Schiller, F. C. S. *The Riddles of the Sphinx: A Study in the Philosophy of Evolution*. London: Swan Sonnenschein, 1891.
———. 'Non-Euclidean Geometry and the Kantian a Priori'. *The Philosophical Review* (March 1896): 173–80.
———. *Studies in Humanism*. New York: Macmillan, 1907.
Schuster, Arthur. 'Letters to the Editor. On Röntgen's Rays'. *Nature* 53 (23 January 1896): 268.
Shrimpton, Nicholas. 'Ruskin and the Aesthetes'. In *Ruskin and the Dawn of the Modern*, edited by D. Birch, 131–52. Oxford: Oxford University Press, 1999.
Silver, Daniel S. 'Knot Theory's Odd Origins'. *American Scientist* 94, no. 22 (2006): 158–65.
Simpson, Anne B. 'The "Tangible Antagonist": H. G. Wells and the Discourse of Otherness'. *Extrapolation* 31, no. 2 (1990): 134–47.
Sirabian, Robert. 'The Conception of Science in H. G. Wells's *The Invisible Man*'. *Papers on Language and Literature* 37, no. 4 (2001): 382–403.
Siraganian, Lisa. 'Out of Air: Theorizing the Art Object in Gertrude Stein and Wyndham Lewis'. *Modernism/Modernity* 10, no. 4 (2003): 657–76.
Smith, Jonathan. *Fact and Feeling: Baconian Science and the Nineteenth-Century Literary Imagination*. Madison and London: University of Wisconsin Press, 1994.
Smith, Jonathan, Lawrence I. Berkove and Gerald A. Baker. 'A Grammar of Dissent: "Flatland," Newman, and the Theology of Probability'. *Victorian Studies* 39, no. 2 (1996): 129–50.
Smith, Lindsay. *Victorian Photography, Painting and Poetry: The Enigma of Visibility in Ruskin, Morris and the Pre-Raphaelites*. Cambridge: Cambridge University Press, 1995.
Sonnenschein, William. *Correspondence with Charles Howard Hinton*, Publishers' Archives, MSS 382, 'Records of George Allen & Unwin, Ltd', Archive and Manuscript Division, Special Collections, University of Reading.
Staubermann, Klaus B. 'Tying the Knot: Skill Judgement and Authority in the 1870s Leipzig Spiritistic Experiments'. *British Journal for the History of Science* 34, no. 1 (2001): 67–79.
Stetz, Margaret Diane. 'Visible and Invisible Ills: H. G. Wells's "Scientific Romances" as Social Criticism'. *Victorians Institute Journal* 19 (1991): 1–24.
Stevens, Hugh. 'Homoeroticism, Identity, and Agency in James's Late Tales'. In *Enacting History in Henry James: Narrative, Power and Politics*, edited by G. Buelens, 126–47. Cambridge: Cambridge University Press, 1997.
Stewart, Balfour and Peter Guthrie Tait. *The Unseen Universe; or, Physical Speculations on a Future State*. London: Macmillan, 1876.
Stromberg, Wayne H. 'Helmholtz and Zoellner: Nineteenth-Century Empiricism, Spiritism, and the Theory of Space Perception'. *Journal of the History of Behavioral Sciences* 25 (October 1989): 371–83.
Summers, Anne. 'The Correspondents of Havelock Ellis'. *History Workshop: A Journal of Socialist and Feminist Historians* 32 (Autumn 1991): 167–83.
Sun, Hon. 'Pound's Quest for Confucian Ideals: The Chinese History Cantos'. In *Ezra Pound and China*, edited by Z. Qian, 96–119. Ann Arbor: University of Michigan Press, 2003.
Sylvester, J. J. 'A Plea for the Mathematician'. *Nature* (30 December 1869): 237–39.
———. 'A Plea for the Mathematician, Part II'. *Nature* (6 January 1870): 261–63.
Taylor, Alexander L. *The White Knight: A Study of C. L. Dodgson*. Edinburgh and London: Oliver and Boyd, 1952.
'The Mystery of the Fourth Dimension of Space'. *Dublin University Magazine*, 1878.

Throesch, Elizabeth. '"The Difference between Science and Imagination?" (Un)framing the Woman in Charles Howard Hinton's "Stella"'. *Phoebe: Journal of Gender & Cultural Critique* 18, no. 1 (2006): 75–98.

———. 'Nonsense in the Fourth Dimension: The "New" Mathematics and the *Alice* Books'. In *Alice Beyond Wonderland: Essays for the Twenty-First Century*, edited by C. Hollingsworth, 37–52. Iowa City: University of Iowa Press, 2009.

Todorov, Tzvetan. *The Poetics of Prose*. Translated by R. Howard. Oxford: Basil Blackwell, 1977.

Tovinio, Stacey A. 'Imaging Body Structure and Mapping Brain Function: A Historical Approach'. *American Journal of Law & Medicine* 33, nos. 2 and 3 (2007): 193–228.

Toynbee, Gertrude, ed. *Reminiscences and Letters of Joseph and Arnold Toynbee*. London: Henry J. Glaisher, 1911.

Traill, H. D. *The New Fiction, and Other Essays*. London: Hurst and Blackett, 1897.

Tucker, Robert. 'Review of "Flatland"'. *Nature* (27 November 1884): 76–77.

Tyndall, John. *Fragments of Science: A Series of Detached Essays, Lectures and Reviews*. London: Longmans, Green and Co., 1872.

Valente, K. G., 'Transgression and Transcendence: *Flatland* as a Response to "A New Philosophy"'. *Nineteenth-Century Contexts* 26, no. 1 (2004): 61–77.

———. '"Who Will Explain the Explanation?": The Ambivalent Reception of Higher Dimensional Space in the British Spiritualist Press, 1875–1900'. *Victorian Periodicals Review* 41, no. 2 (Summer 2008): 124–49.

Ward, J. A. *The Search for Form: Studies in the Structure of James's Fiction*. Chapel Hill: University of North Carolina Press, 1967.

———. '*The Ambassadors* as a Conversion Experience'. *Southern Review* 5, no. 2 (1969): 350–74.

Waters, Isobel. '"Still and Still Moving": The House as Time Machine in Henry James's *The Sense of the Past*'. *Henry James Review* 30, no. 2 (2009): 180–95.

Wells, H. G. 'Scepticism of the Instrument'. *Mind* 13, no. 51 (1904): 379–93.

———. *The Wonderful Visit*. London: J. M. Dent and Sons, 1914.

———. *Boon, the Mind of the Race, the Wild Asses of the Devil, and, the Last Trump: Being a First Selection from the Literary Remains of George Boon*. London: T. Fisher Unwin, 1920.

———. *The Scientific Romances of H. G. Wells*. London: Victor Gollancz, 1933.

———. *Experiment in Autobiography: Discoveries and Conclusions of a Very Ordinary Brain, Since 1866*. London: Victor Gollancz and Cresset Press, 1934.

———. *The Definitive Time Machine: A Critical Edition of H. G. Wells's Scientific Romance*. Edited by H. M. Geduld. Bloomington: Indiana University Press, 1987.

———. *The Complete Short Stories of H. G. Wells*. Edited by J. R. Hammond. London: J. M. Dent, 1998.

———. *A Modern Utopia*. London: Penguin, 2005.

———. *The Invisible Man: A Grotesque Romance*. London: Penguin, 2005.

Whewell, William. *The Philosophy of the Inductive Sciences*. London: John W. Parker, 1847.

Wierzbicki, James Eugene. *Film Music: A History*. New York: Routledge, 2009.

Wilde, Oscar. *Complete Works*. Edited by V. Holland. London and Glasgow: Collins, 1966.

Williams, Keith. *H. G. Wells, Modernity and the Movies*. Liverpool: Liverpool University Press, 2007.

Williams, Merle A. *Henry James and the Philosophical Novel: Being and Seeing*. Cambridge: Cambridge University Press, 1993.

Williams, Raymond. *The Long Revolution*. Orchard Park, NY: Broadview Press, 2001.

Wise, M. Norton. 'Time Discovered and Time Gendered in Victorian Science and Culture'. In *From Energy to Information: Representation in Science and Technology, Art, and Literature*, edited by B. Clarke and L. D. Henderson, 39–58. Stanford: Stanford University Press, 2002.

Woolf, Virginia. *The Captain's Death-Bed and Other Essays*. London: Hogarth Press, 1950.

'X Ray Photography'. *Scientific American* (March 7, 1896): 115.

Zamir, Shamoon. *Dark Voices: W. E. B. Du Bois and American Thought, 1888–1903*. Chicago and London: University of Chicago Press, 1995.

Zöllner, J. C. Friedrich. 'On Space of Four Dimensions'. *Quarterly Journal of Science* 8 (April 1878): 227–37.

——. *Transcendental Physics: An Account of Experimental Investigations*. Translated by C. C. Massey. London: Harrison, 1880.

Zunshine, Lisa. *Why We Read Fiction: Theory of Mind and the Novel*. Columbus: Ohio State University Press, 2006.

INDEX

Abbott, Edwin A.
   Flatland 2, 15, 24–26, 46, 152, 163, 174, 181
Aesthetic movement 8, 13, 14, 39, 85
Anschütz, Ottomar 100
Ansell-Pearson, Keith 94, 103, 104
Auden, W. H. 181

Bain, Alexander 26, 58
Ballard, Marvin H. 110
Barthes, Roland 45–46
Batchelor, John 14, 156
Baudrillard, Jean 24, 152, 161
Beaumont, Matthew 78
Beer, Gillian 8, 19, 26, 50, 66, 146
Bell, Ian F. A. 41, 48, 49, 68
Bell, Millicent 174
Bentham, Jeremy 59
Bergson, Henri 137, 195
   la durée pure 68, 196
   and time 196
Berkeley, George 53
Bigelow, William Sturgis 110
Bold, Paul 24
Bólyai, Johannes 22
Boole, George 3, 79
Boole, Mary Everest 79
Borges, Jorge Luis 2–3, 197
Bradbury, Nicola 172
Bragdon, Claude 41, 151
Brooke, Emma 79
Brown, Bill 175

Carpenter, Edward 10, 76, 87, 88
   From *Adam's Peak to Elephanta* 80–81
Carroll, Lewis *See* Dodgson, Charles L.
Cartesian dualism 70, 108
Cartwright, Lisa 145, 148, 149, 160
Caso, Pere Borrell del 154
Cassirer, Ernst 63, 65
Cayley, Arthur 21, 23
Chapman, Sara 183
Clarke, Bruce 44, 52, 53, 55, 63, 64, 75, 174
Clifford, W. K. 22, 24
Conrad, Joseph 2, 4
   Inheritors 81, 168
Cubism 160
   and fourth dimension 15, 71
Culverhouse, Emily 149

Daley, Kenneth 39
Daughtery, Sarah B. 164
Derrida, Jacques 66
Devlin, Keith 5
Dewey, John 111
dimensional analogy 12, 15, 20, 24–31, 35, 43, 46, 49, 51, 53, 86, 118, 119, 134, 135, 163, 168, 169, 172, 174, 187, 196
Dodgson, Charles Lutwidge 8, 127
Donadio, Stephen 187
Du Bois, W. E. B. 2, 4, 113
   'Vacation Unique' 113n30
Du Maurier, George 168

Edel, Leon 191
Edison, Thomas 100
Einstein, Albert 1, 25n26, 195
Eliot, T. S. 100
Ellis, Edith 91
Ellis, Havelock 10, 91
   and Hinton's bigamy trial 4, 80
   and James Hinton 3, 43, 76, 78, 79
   and Olive Schreiner 77, 80
   on sex and gender 82, 140
Emerson, Ralph Waldo 162, 169
entropy 52, 54, 56, 57, 63, 75, 76
eternal recurrence 10, 91, 92, 94, 96, 97–103
Euclid 19, 20, 21, 22
evolution 54

Fabian Society 77, 78, 79, 80, 89
Fechner, Gustav Theodor 126
   'Der Raum hat Vier Dimensionen' 24, 26–27
   Elemente der Psychophysik 127
   'mother-sea' metaphor 10, 43, 108, 126–28, 130, 195
   Zend-Avesta 129
Fellows, Jay 38

Fellowship of the New Life 77, 78, 79, 80, 89
Fenollosa, Ernest 110
flat vs round characters 7, 160, 174
Florence, Maude 3, 44, 84
Fogel, Daniel Mark 169, 188, 189
Ford, Ford Madox 4
   character in *Boon* 161
   on 'Great Good Place' 178, 179
   Inheritors 81, 168
Forster, E. M. 7, 160, 174
fourth dimension. *See also* hyperspace philosophy
   conception of 37, 48, 49, 52, 72, 74
   in W. E. B. Du Bois 113n30
   in Edward Carpenter 80–82
   in Henry James 1, 12, 177, 178, 180, 184
   and 'higher consciousness' 7, 13, 39, 87, 94, 121, 122, 131, 162, 165, 194, 197
   and hyperbeing 53, 66, 67, 117, 123, 174
   and hypostasization 23, 26, 53
   as interpretive lens 8, 10, 11, 13, 109, 164, 167, 193
   and Modernism 4, 41, 83
   and Newtonian physics 33
   and non-Euclidean geometry 8, 16, 21, 22
   perception of 32, 48, 52, 74
   and 'queering' 11, 138
   series of slices 5, 8, 51, 52, 68, 72, 136, 160, 190
   and Spiritualism 4, 30
   in *Unseen Universe* 56, 62, 153
   in H. G. Wells 2, 11, 13, 15, 133, 134, 136, 138, 140
   in William James 113, 124
Frank, Joseph 100
Freud, Sigmund 10, 81, 123, 131, 149

Gauss, Carl Friedrich 23, 27
Gelley, Alexander 172
Gibbons, Tom 133
Glasser, Otto 150
Gooding-Williams, Robert 98
Gosse, Edmund 161
Green, Thomas Hill 8, 35
Grosz, Elizabeth 83

Haddon, Caroline 78, 79
Hall, G. Stanley 116, 127
Hayles, N. Katherine 50, 56, 71
Hearn, Lafcadio 110n12
Helmholtz, Hermann von 22, 28, 30, 31, 33

Helsinger, Elizabeth K. 36, 185
Henderson, Andrea 52
Henderson, Linda Dalrymple 2, 20, 27, 147
Herbert, Christopher 4
Hilton, Tim 85
Hinton (*née* Boole)
   Mary Ellen 3, 69, 111, 129
Hinton, Charles Howard
   bigamy 3–4, 9, 43–44, 77, 78, 112
   on British imperialism 85
   'Casting Out the Self' 8, 45, 46, 67–74, 117, 123, 136
   correspondence with William James 10, 87, 92, 107, 109–13, 118, 131
   cube exercises 9, 46, 67, 68, 72, 79n21, 115, 126, 136, 160, 195
   'Education of the Imagination' 76
   The Fourth Dimension 2, 113, 118, 125, 195
   influence on H. G. Wells 134, 136, 137, 140–45
   in Japan 3, 80, 110, 110n12
   'Language of Space' 113, 195
   'Many Dimensions' 76, 103
   'mother-sea' metaphor 130–31
   New Era of Thought 19, 44, 73, 77, 84, 115, 117
   at Oxford 8, 32, 35, 85
   'Persian King' 8, 47, 53, 54, 57–67, 75, 76, 103, 129, 169, 174, 176
   'Picture of our Universe' 46
   'Plane World' 46, 138–40, 151
   Preface to *Stella* and *Unfinished Communication* 104
   Science Note-book 3, 80
   on sex and gender 84, 88, 89, 138
   on 'sexual inversion' 11, 138–40
   Stella 9, 10, 36, 76, 83–91, 104, 113, 137, 140–42, 146, 148, 155
   and time 10, 52, 83, 87, 91, 92, 100, 103, 117, 134, 173n27, 184, 190, 196
   Unfinished Communication 9, 10, 76, 83, 92, 95–97, 99–104, 112, 113, 118, 121, 122, 129, 130, 135, 187
   'What is the Fourth Dimension?' 8, 30–31, 48–51, 65, 66, 87, 113, 117, 129, 150, 155
Hinton, James 3, 10, 43, 78, 91
   Chapters on the Art of Thinking 3, 32, 110
   correspondence with Charles Howard Hinton 40, 42
Hintonians 79–80

INDEX   211

influence on Charles Howard Hinton 20,
    43, 50, 60, 84
influence on Havelock Ellis 77, 79
and John Ruskin 32
'law breaking' 32, 50, 60, 76
'service' 32, 42–44, 60, 76, 79
on sex and gender 79, 82
sexual scandal 3, 79–80
Hinton, Margaret 78, 79
Hocks, Richard A. 7, 186, 187
Hodgson, Shadworth 110
Holbein, Hans the Younger 185
Hollingdale, R. J. 95
Howells, W. D. 9
Hutchison, Hazel 11, 167, 172
Huxley, T. H. 20
hyperreal 24, 165, 187, 193, 196
hyperspace philosophy 1, 29, 35, *See also*
    fourth dimension
    and aesthetics 2, 9, 40
    and British socialism 78, 80, 81, 83, 89,
        90, 168
    and ethics 28, 49, 74, 75, 78, 80, 81,
        94, 113
    and Henri Bergson 68, 196
    and Henry James 11, 167
    and pragmatism 108
    and 'transcendental materialism' 40, 41,
        42, 45, 196

Ibsen, Henrik 19, 92
Iser, Wolfgang 9, 47, 72, 121

Jackson, Holbrook 19
James, Henry 6, 7, 160
    Ambassadors 168, 170–73, 188
    Awkward Age 169, 170
    'Beast in the Jungle' 189
    character in Boon 161, 163
    'Great Good Place' 12, 169, 177–84
    Golden Bowl 11–12, 182
    'House of Fiction' 130, 191
    Jamesian central consciousness 11, 16, 125,
        163, 165, 167, 168, 169, 173, 176, 184,
        185, 187, 188, 193
    'Jolly Corner' 13, 169, 185–93
    Princess Cassamassima 15, 182
    Sense of the Past 190
    Spoils of Poynton 1, 12, 169, 173–77, 179
    theory of art 6, 11, 165, 183, 193
    and time 180, 184, 190

Turn of the Screw 164
Wings of the Dove 184, 188
James, William 4, 6, 40, 168, 191
    ambulatory relations 7, 8, 116, 120,
        129, 175
    on attention 121–22, 125
    correspondence with Charles Howard
        Hinton 10, 87, 92, 107, 109–13, 118, 131
    correspondence with H. G. Wells 134
    on fourth dimension 113, 113n30, 195
    and free will 122–24
    on Henri Bergson 195
    'Human Immortality' 127
    on Immanuel Kant 116
    'mother-sea' metaphor 108,
        126–28, 130–31
    Pluralistic Universe 195
    Pragmatism 6
    Principles of Psychology 114, 116, 119, 123
    stream of consciousness 92, 107, 109, 114–
        15, 118, 119–21
    'Will to Believe' 16, 113, 124, 125
Jann, Rosemary 2
Jenkins, Alice 8, 13, 20, 51
Jordanova, Ludmilla 144
Jowett, Benjamin 3
Joyce, James 47, 100

Kant, Immanuel 8
    Critique of Pure Reason 35
    idealism 28, 29, 34
    influence on Charles Howard Hinton 20,
        28, 35, 52, 124
    and space 28, 33, 34, 35
Kern, Stephen 64
Koven, Seth 77

Land, Jan Pieter Nicholas 33–35, 38
Lawrence, D. H. 4, 12, 192
Levin, Jonathan 108, 112, 125, 170, 184, 193
Levine, George 39, 67, 68, 73
Levy, Oscar 95
Lewes, George Henry 21, 29, 30
Lland, Meriel 41
Lobachevskii, Nicholai 22
Lumiere, Louis 100
Lustig, Timothy 16, 173

Mallarmé, Stéphane 197
Mallock, W. H. 161
Mann, Thomas 147

Marcus, Laura 145
Marshall, M. E. 126, 127
Maxwell, James Clerk
  Maxwell's demon 55, 56, 63, 103
McDonald, Henry 193
McGurl, Mark 7, 12, 15, 41, 163, 174, 182
McMurray, William 180
Men and Women's Club 78, 79, 80
Metaphysical Society 3, 32
Meyer, Steven 164
Michie, Helena 142
Minkowski, Hermann 3
Modernism 4, 19, 20, 48, 64, 83, 100, 108
  and Henry James 167, 168
  and William James 108
Moore, George 161
Mowbray, J. P. 184, 185, 188
Müller, Max 63
Murphy, Patricia 75, 82
Muybridge, Eadweard 100

Newcomb, Simon 111n21
Newman, Ernest 15, 138
Nietzsche, Friedrich 10, 32, 91, 103, 162
  'Prelude to a Philosophy of the Future' 104
  on sex and gender 104
  Thus Spake Zarathustra 10, 93, 95, 97–99
  and time 93, 94, 97–99
  'Uses and Disadvantages of History' 92, 95, 96
non-Euclidean geometry 8, 19, 20, 22–23, 29, 35, 53
Nordau, Max 14, 92

O'Gorman, Francis 36
Oakeshott, Joseph 78

Papin, Lilianne 33
Pater, Walter 10, 39, 40, 41, 73
pathetic fallacy 36, 39, 62, 63
Peard, George 83
Pearson, Karl 4, 77, 78, 205
Peirce, Charles Sanders 111
Plato's cave allegory 15
Potts, Annie 142
Poulet, Georges 184
Pound, Ezra 2, 19, 100
pragmatism 6, 108
Proust, Marcel 2

Rampley, Matthew 10, 94, 97, 98, 99
Rawlings, Peter 190
Raynal, Maurice 71
Richards, Joan 23
Richardson, Joan 108, 168, 185
Röntgen, Bertha 149
Röntgen, Wilhelm 10, 133, 147, 150
Ruskin, John 13, 46, 84
  on aesthesis and theoria 39, 185
  first order of poets 37, 44, 63
  Fors Clavigera 32, 36
  Hinksey road project 8, 35, 85
  and imagination 31, 37–39, 47, 50, 52, 68, 72, 121, 126
  influence on Charles Howard Hinton 20, 35–36, 39
  and James Hinton 32
  Modern Painters 36, 37
  paradoxical invisible man 38, 42, 67, 90, 159
Rylance, Rick 26, 29

Saiber, Arielle 5
Salmon, George 23
Salmon, Richard 173, 182
Scheick, William J. 11, 71, 133, 151, 153, 170
Schiller, F. C. S. 6, 28, 110, 122
Schreiner, Olive 4, 76, 77, 80, 83, 88
  and Hinton's bigamy trial 4, 80
Schuster, Arthur 148
scientific romance 26, 30, 47, 50, 60, 67, 72, 137, 157
Shrimpton, Nicholas 39
Sidgwick, Henry 43
simulacrum 24, 152, 196
Sirabian, Robert 156
Slade, Henry 30
Smith, Jonathan 22, 33
Smith, Lindsay 38
Society for Psychical Research 16, 128
Sonnenschein, William 77, 146
Spencer, Herbert 21, 26
spirals
  in Charles Howard Hinton 118, 119
  pattern in Henry James 169–70, 171, 189, 193
  pattern in H. G. Wells 170
Stein, Gertrude 2, 164
Stevens, Hugh 189
Stewart, Balfour 54–57, 62, 66, 75, 153

Stumpf, Carl 116
Sylvester, James Joseph 21–22, 23, 27, 35

Tait, Peter G. 54–57, 62, 66, 75, 153
Taylor, Alexander L. 27
tesseract 22, 126
Theosophy 30
thermodynamics 58, 63, 146
    second law of 53–54, 55, 57, 59, 60, 75
Thomson, William 54, 75
Thring, Edward 3
Todorov, Tzvetan 175
Toynbee, Arnold 35
Traill, H. D. 19
trompe-l'oeil 152, 160
Turner, Henry S. 5
Tyndall, John 54, 62

Übermensch 92, 94, 162
unlearning 95–97, 129

Valente, K. G. 22

Waltershausen, Sartorius von 27
Ward, J. A. 170
Wells, H. G. 2, 10, 168
    Boon 11, 154, 160–63, 168
    'Chronic Argonauts' 134, 135–36
    'Cyclic Delusion' 170
    *Experiment in Au*tobiography 163
    four-dimensional aesthetic 134, 151, 153, 154, 157, 164, 165, 169
    Invisible Man 10–11, 84, 136, 137, 140–45, 151, 154, 155–59
    Modern Utopia 134, 154, 163, 164
    'Plattner Story' 140
    quarrel with Henry James 11, 39, 156, 160, 161, 163, 165

splintering frame technique 11, 133, 151–54, 160, 163, 164, 168, 169, 170, 193
'Through a Window' 153–54, 172
and time 134, 135
Time Machine 13, 135, 136–37
Wonderful Visit 13–15, 138, 143, 159
Whale, James 145
Whewell, William 21, 33
Wilde, Oscar 2, 14, 15, 85, 88, 92, 138
will 6, 97
    in Charles Howard Hinton 64, 97, 102, 103, 125
    in Friedrich Nietzsche 93, 95, 98
    in William James 16, 114, 122–24
    will zur macht 95
    will-to-attention 125
    will-to-create 10, 92, 95
    will-to-passivity 62n70, 63, 66, 95
    will-to-power 10, 62n70, 183
Williams, Keith 101, 135, 137, 145
Williams, Merle A. 175
Williams, Raymond 49
Wise, M. Norton 75
Woolf, Virginia 4, 10, 152

X-rays 5, 145, 159
    in Culverhouse's 'Photography' 149–50
    discovery of 11, 133, 146–48, 149
    and fourth dimension 11, 15, 43, 133, 137, 150, 151
    in *The Invisible Man* 145–46, 151, 154
    in *Magic Mountain* 147, 148
    and occult 150
    and *Stella* 11, 146

Zamir, Shamoon 113
Zöllner, J. C. Friedrich 30, 125

www.ingramcontent.com/pod-product-compliance
Lightning Source LLC
Chambersburg PA
CBHW021827300426
44114CB00009BA/350